M4

Water Fluoridation Principles & Practices

Sixth Edition

American Water Works Association

Manual of Water Supply Practices—M4

Water Fluoridation Principles & Practices

Managing Editor: Melissa Valentine
Production: Studio Text

If you find errors in this manual, please email books@awwa.org. Possible errata will be posted at www.awwa.org/resources-tools/resource.development.groups/manuals-program.aspx.

Library of Congress Cataloging-in-Publication Data

Names: Copeland, Ari, author. | American Water Works Association, issuing body.
Title: Water fluoridation principles and practices / by Ari Copeland.
Other titles: AWWA manual ; M4.
Description: Sixth edition. | Denver, CO : American Water Works Association, [2016] | Series: Manual of water supply ; M4
Identifiers: LCCN 2016019717 | ISBN 9781625761705
Subjects: LCSH: Water--Fluoridation.
Classification: LCC TD467 .C66 2016 | DDC 628.1/663--dc23 LC record available at https://lccn.loc.gov/2016019717

ISBN-13 978-1-62576-170-5
eISBN-13 978-1-61300-383-1

Printed in the United States of America
American Water Works Association
6666 West Quincy Avenue
Denver, CO 80235-3098
awwa.org

Printed on recycled paper

Contents

Figures

This page intentionally blank.

Tables

This page intentionally blank.

Preface

The American Water Works Association has been an active partner in water fluoridation from the original pioneering studies on its implementation and effectiveness from 1945 through 1951, and has supported community water fluoridation since 1951.

In addition, from the 1940s to the 1960s, the US Public Health Service provided national leadership in water fluoridation practice and issued periodic technical advisories on water fluoridation. The US Environmental Protection Agency (USEPA) provided support and guidance on technical practice from 1972 to 1978 until the Centers for Disease Control and Prevention (CDC) assumed responsibility for providing support to state water fluoridation programs in 1978.

These organizations' wealth of knowledge gained through the experiences and management of several thousands of fluoridation process installations can improve future installations and enhance successful continued operations to provide improved health for our communities.

This manual is a resource to assist decision makers planning to use fluoridation treatment, engineers designing and installing these facilities, and water utility personnel managing water operations. The manual presents guidelines and is not intended to take the place of expert advice. Anyone planning or using fluoridation should carefully consider fluoride research, regulations, and methods. State or provincial regulatory requirements should always be the first point of reference for water fluoridation design and practices to improve the health of the citizens.

The first edition of the American Water Works Association M4 *Water Fluoridation Principles & Practices* was prepared from material supplied and previously published by the USEPA. The second edition was updated and revised using additional technical collaboration with the US Centers for Disease Control and Prevention. The third, fourth, and fifth editions were updated and revised incrementally by section. This sixth edition represents a substantial updating and revising of the manual's content to remain consistent with industry practices and to reflect changes in regulatory and public health guidance and the experience of contributing authors.

This page intentionally blank.

Acknowledgments

This manual was revised by the AWWA Treatment Plant Operations and Maintenance Committee.

The following committee members served as the editorial review board in the preparation of this edition of the manual:

Ari Copeland, Chair, Black & Veatch Corporation, Waterbury, Conn.
David Baker, Enprotec / Hibbs & Todd, Inc., Colorado Springs, Colo.
Angela Kana-Spitz, Short Elliott Hendrickson Inc., Denver, Colo.
Barbara Melcher, CDM Smith, Carlsbad, Calif.
Michael Wentink, Nebraska DHHS, Division of Public Health, Office of Drinking Water and Environmental Health, North Platte, Neb.

Additional contributing committee members, M4 *Water Fluoridation Principles & Practices*:

Dr. Katherine Ann Alfredo, Staten Island, N.Y.
Charly C. Angadicheril, City of Fort Worth, Fort Worth, Texas
John A. Consolvo, City of Philadelphia, Philadelphia, Penn.
Stephen Gasteyer, Michigan State University, East Lansing, Mich.
Michael S. Grimm, Happy Valley, Ore.
Kimberly Gupta, Portland Water Bureau, Portland, Ore.
David Heumann, Los Angeles Department of Water and Power, Los Angeles, Calif.
Daniel Huggins, City of London Water Operations, London, Ont., Canada
Dr. Alisha Knowles, Halifax Water, Halifax, N.S., Canada
Hoy Yi (Mandy) Lai, CDM Smith, Walnut Creek, Calif.
Brian John Leto, Pasadena, Calif.
Edward A. Moreno, DDS, West Liberty, Iowa
Tom Napier, Dept. of State Health Services, Austin, Tex.
Bryan R. Phinney, Keller Associates, Inc., Pocatello, Idaho
Chet V. Shastri, City of Fort Wayne, Fort Wayne, Ind.
Preeti Shridhar, City of Renton, Renton, Wash.
Steve H. Via, AWWA, Washington, D.C.
Dr. June M. Weintraub, San Francisco Dept. of Public Health, San Francisco, Calif.
Jason Yoshimura, CDM, Carlsbad, Calif.

Molly Beach, former Manuals Specialist, Mindy Burke, current Manuals Specialist, and Steven J. Posavec, Standard Methods Manager, provided AWWA staff support.

The editorial review committee would like to acknowledge the contribution of Kip Duchon, National Fluoridation Engineer for the Centers for Disease Control and Prevention, for his assistance during the editing process.

This page intentionally blank.

Chapter 1

Water Fluoridation

INTRODUCTION

The goal of this manual is to assist with the planning and operation of fluoridation systems by decision makers, design engineers, and water utility personnel. This chapter discusses fluoride occurrence, growth of community water fluoridation, and legal issues surrounding fluoridation. The regulatory requirements of community water fluoridation are also addressed, including both federal regulations and the varying approaches states have used to implement fluoridation programs. Additionally, fluoridation outside of the United States is discussed.

Fluoridation in this manual refers to the addition of fluoride to drinking water to maintain a recommended level to improve oral health. Fluoridation was named as one of the Ten Great Public Health Achievements in the 20th Century by the Centers for Disease Control and Prevention (CDC) along with the use of chlorine for disinfection of public water supplies (CDC Morbidity and Mortality Weekly Report, April 2, 1999). Control of infectious diseases has resulted from clean water and improved sanitation. Infections such as typhoid and cholera transmitted by contaminated water, a major cause of illness and death early in the 20th century, have been reduced dramatically by improved sanitation. Water fluoridation was first implemented in 1945, and in 1951, the National Research Council (NRC) of the National Academy of Sciences (NAS), the US Surgeon General, and professional organizations including the American Water Works Association (AWWA) and the American Dental Association (ADA) recommended that communities implement water fluoridation. The US Public Health Service (USPHS) recommended a range of 0.7 to 1.2 mg/L (based on annual average ambient temperature) as part of the 1962 Drinking Water Standards. In 2011, the US Department of Health and Human Services (USHHS) proposed changing the recommended fluoride level in drinking water to a single value of 0.7 mg/L. According to national health surveillance statistics reported by the USPHS and the CDC, the number of people with access to fluoridated water continues to increase and in 2012, 210.6 million people in the United States had access to fluoridated water.*

* 2012 Water Fluoridation Statistics from the US Centers for Disease Control (CDC).

OCCURRENCE

Fluorine, a gaseous halogen, is the 13th most abundant element in the earth's crust. Fluorine is also the most electronegative element, so it is not found in its free elemental form in nature. Instead, it exists as a mineral such as fluorosilicates in granites, calcium fluorides in some ores, along with other mineral forms, or as a dissolved reduced ionic form in solution. Fluoride ion solubility varies, but the sodium and potassium salts of fluoride are highly water soluble.

There are both natural and anthropogenic sources of fluoride. Natural sources of fluoride include volcanic and geothermal activity emissions, weathering of certain types of rocks, wind-blown erosion of soils, and marine origin. Commercial ore deposits of fluorspar, fluorapatite, and cryolite are the source for most fluoride products. Anthropogenic releases of fluoride occur predominantly because of the burning of coal for power production and are also caused by manufacturing, steel and aluminum production, and oil refining.

Fluoride is ubiquitous in the environment and therefore likely to be present to some extent in all water sources. Seawater contains approximately 1.2 to 1.5 mg/L of fluoride. The concentration present in source water is often equal to the fluoride in rainfall, which is typically 0.1 to 0.2 mg/L. However, natural levels in proximity to volcanic sources can be significantly higher because volcanic emissions are recognized as a significant environmental source. The naturally occurring concentrations in surface waters are generally lower than those needed to promote good dental health. Fluoride concentrations in groundwater can be more variable than in surface water, with the concentration dependent on the geological setting. In the United States, groundwater levels typically range from nondetectable levels to greater than 4 mg/L. However, elevated levels of naturally occurring fluoride are not common. Based on reported natural fluoride levels in community public water systems, the CDC estimates that less than 0.5% of the US population served by public water systems has natural fluoride levels exceeding 2 mg/L, and less than 0.1% of the population served by public water supplies has naturally occurring fluoride in excess of 4 mg/L. The US Geological Survey (USGS), in a survey of private groundwater wells, estimated that up to 4% of private wells exceed 2 mg/L and up to 1.2% exceed 4 mg/L (Quality of Water from Domestic Wells in the United States, Leslie DeSimone et al. November 2009). Private wells can have a higher incidence of elevated fluoride as many homeowners do not test their wells for contaminants, while public water supplies are required to report on contaminants levels and have worked to identify alternate sources when possible.

HISTORY OF USE (GROWTH OF COMMUNITY WATER FLUORIDATION)

The importance of fluoride as a nutrient for good oral health resulted from environmental observations of communities with naturally occurring fluoride. In the 1920s and 1930s, fluoride was identified as a factor in tooth development. Investigations showed that in communities having naturally fluoridated water, fluorosis (mottled or discolored teeth) occurred when the fluoride content of the water was abnormally high. However, these teeth also showed decreased incidence of tooth decay or dental caries, which was unusual considering the large prevalence of tooth decay in that era. During the 1930s, studies concluded that the beneficial range of fluoride to prohibit dental caries while still minimizing risk of dental fluorosis was 1.0 mg/L to 1.5 mg/L, with an optimum concentration of 1.0 mg/L to achieve the beneficial level while remaining below the higher levels that could stain teeth.

Starting in 1945, initial community trials were held to test the theories that fluoride reduced tooth decay formulated over the preceding decades. In each of these trials, one city implemented community water fluoridation, while at least one other city acted as the control. The cities had parallel characteristics except for the fluoride content of the water supply. Several of these studies, implemented between 1945 and 1947, are listed here (naturally occurring fluoride levels are given in parentheses):

1. Grand Rapids, Mich. (0.15 mg/L): Adjusted to 1.0 mg/L as part of this study. This city was compared with the control cities of Muskegon, Mich. (0.15 mg/L) and Aurora, Ill. (naturally 1.2 mg/L).
2. Newburgh, N.Y. (0.1 mg/L): Adjusted to 1.1 mg/L as part of this study. The control city for this trial was Kingston, N.Y. (0.1 mg/L).
3. Evanston, Ill.: Adjusted to approximately 1 mg/L as part of this study. This city was compared with the control city of Oak Park, Ill. (negligible amounts of natural fluoride).
4. Brantford, Ont. (0.1 mg/L): Adjusted to 1.0 mg/L as part of the study. This city was compared with the control cities of Sarnia, Ont. (0.1 mg/L), and Stratford, Ont. (naturally 1.2 mg/L).

On Jan. 25, 1945, Grand Rapids, Mich., became the first city to add fluoride to its water supply. In this study, the oral health in Grand Rapids was compared with that of Muskegon, a nearby community consuming water from the same source without fluoride addition. Six years after the study began, surveys indicated that tooth decay levels in six-year-old children (i.e., those born since fluoridation commenced) in Grand Rapids was approximately half that of Muskegon.

At the time of these early community trials, poor oral health and widespread dental decay was common, so favorable results were met with great interest. The analysis of the first results from these community trials was published by the National Research Council (NRC) of the National Academy of Sciences (NAS) in November 1951 and summarized in the January 1952 issue of *Journal AWWA* (Vol. 44, no. 1, 1–8) with the recommendation that all cities with a child population of sufficient size should fluoridate the water supply. Preliminary reports from the NRC in 1951 prompted the US Surgeon General to recommend that cities should fluoridate their water supplies. In July 1951, city officials in Muskegon decided to withdraw from the study and fluoridate the city's water supply. Similar results were observed in the other trials referenced, i.e., significant reduction in dental decay rates was observed in the cities that were fluoridating, with little or no change in the controls. Since that time, the US population receiving optimally fluoridated drinking water has continued to increase (Figure 1-1).

In 1951, major organizations such as the USPHS, NRC, US Surgeon General, and AWWA all endorsed community water fluoridation. In 1962, the USPHS recommended that fluoride in drinking water should be 0.7 to 1.2 mg/L based on ambient annual temperature to reflect the different consumption of water between warmer climates and cooler climates. In 2011, the Department of Health and Human Services proposed changing this range to a single recommended value of 0.7 mg/L to reflect modern exposures and water consumption; this is discussed in further detail in chapter 2.

The AWWA policy statement on fluoridation was adopted by the organization's board of directors on Jan. 25, 1976, reaffirmed on Jan. 31, 1982, and revised on Jan. 20, 2002, Jan. 21, 2007, and Jan. 22, 2012. The official text is as follows:

> The American Water Works Association (AWWA) supports the recommendations of the World Health Organization (WHO), American Medical Association (AMA), Canadian Medical Association (CMA), Centers for Disease Control (CDC), American Dental Association (ADA), Canadian Dental Association (CDA), and other professional organizations in the medical community, for the

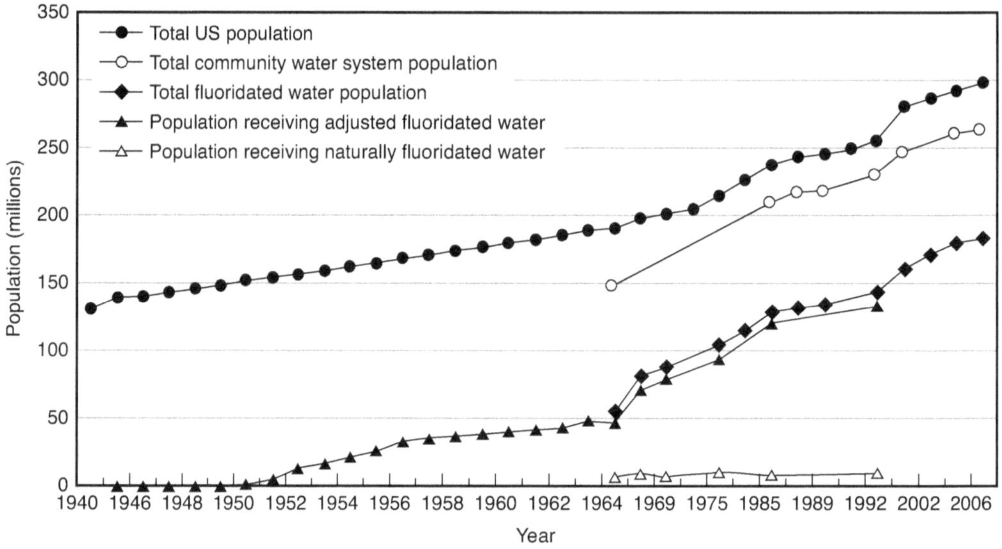

Figure 1-1 Growth of community water fluoridation around the world

Source: CDC - Fluoridation Growth. 2012.

> fluoridation of public water supplies as a public health benefit. AWWA supports the application of fluoride in a responsible, effective, and reliable manner that includes monitoring and control of fluoride levels mandated by provincial, state, and/or federal laws and that is subject to community acceptance through applicable local decision-making processes. AWWA is committed to regular reviews of the most current research on fluoride and the positions of the medical and dental communities.

In 1945, Brantford, Ont., became the first Canadian city to add fluoride to its water as part of the community trials referenced previously. Approximately 45 percent of the Canadian population using public water supplies receives fluoridated water.

As the use of drinking water fluoridation increased in the United States and Canada, countries in Europe, South America, Africa, and Asia also began implementing water fluoridation. By the year 2004, water fluoridation was being practiced in more than 60 countries for almost 405 million people. In addition, another 50 million people have water supplies that are considered naturally fluoridated at levels sufficient for good oral health. Fluoridation is practiced extensively in Australia, Brazil, Canada, Chile, Columbia, Ireland, Israel, Malaysia, New Zealand, People's Republic of China, Hong Kong, Singapore, and the United Kingdom. In addition to US organizations such as the American Medical Association (AMA), American Dental Association (ADA), and the CDC, international organizations such as the World Health Organization (WHO) support water fluoridation.

More detailed information about fluoridation in the international community can be found in the American Dental Association's *Fluoridation Facts* document.

> *Benefits of community water fluoridation.* Fluoridation is a cost-effective and practical form of preventing tooth decay. The benefits of community water fluoridation include prevention of dental caries, the ability to provide a consistent and optimized level of fluoride, and an inexpensive mechanism to provide health benefits to all, regardless of age, income, education, or socioeconomic status. Fluoridation has traditionally been justified based on the economics of a 40 to 65 percent reduction in dental caries. The CDC estimates that every $1 invested

in water fluoridation yields an approximate $38 in savings from dental treatment costs in most cities (Griffin et al. 2001).

Legal Issues

Fluoridation of public water supplies has not been without controversy. Opponents of community water fluoridation have challenged its safety and effectiveness. The legality of fluoridation has been tested in the courts numerous times. Beginning in 1952, injunctions were sought in some communities to prevent the initiation or continuance of fluoridation. These cases were generally based on the following types of arguments:

- Violation of religious freedom
- Abuse of police power (the ability of the state to make laws concerning the health, safety, and welfare of its citizens)
- Infringement of personal choice through mass medication, class legislation (i.e., only children benefit)
- Violation of pure food acts
- An unreasonable or unnecessary measure, wasteful or illegal use of public funds
- An unsafe measure or nuisance
- Availability of alternative
- Breach of contract
- Deprivation of fundamental liberties

In more than 65 years of litigation, fluoridation has withstood challenges based on constitutional objections (related to the First, Ninth, Tenth, Fourteenth, and Fifteenth Amendments). It has withstood legal challenges in more than 25 states and has been upheld in the highest courts in more than a dozen states. The US Supreme Court has denied review several times, generally on the grounds that no substantial federal or constitutional questions were involved. Courts have viewed fluoridation as a proper means of furthering public health and welfare, rejecting the legal argument that fluoridation ordinances are a deprivation of freedoms guaranteed under the Constitution.

More information can be found at www.fluidlaw.org, which is a database maintained by American University's Washington College of Law and the Children's Dental Health Project. FLUID (the Fluoride Legislative User Information Database) is a collection of legal documents related to the fluoridation of water supplies.

REGULATORY

Federal

Fluoridation of public water supplies is not mandated by the US Environmental Protection Agency (USEPA) or any other federal agency in the United States. The 1974 Safe Drinking Water Act (SDWA) specified that no national primary drinking water regulation can require the addition of any substance for preventive health benefits not related to drinking water contamination. This prohibition inherently established fluoridation as a decision to be made by each individual state or local municipality.

The SDWA further required that the USEPA determine the level of contaminants in drinking water at which no adverse health effects are likely to occur, and to apply limits based on possible health risks assuming a lifetime of exposure. These limits (which are nonenforceable) are maximum contaminant levels goals (MCLGs). The enforceable limits are maximum contaminant levels (MCLs). Though MCLs are set as close as possible to MCLGs, the enforceable MCLs must also consider cost, the ability to provide a meaningful impact on human health, and the ability of utilities to detect and remove contaminants. MCLs are never lower than MCLGs, but may be higher.

The Interim Regulation for Fluoride was promulgated in 1975 as a National Interim criteria with the MCL varying from 1.4 to 2.4 mg/L based on annual ambient temperature that was double the level recommended by the USPHS. This Interim Standard was challenged in 1981 by South Carolina with a request to delete the MCL but instead implement a Secondary Maximum Contaminant Level (SMCL) for dental fluorosis. In 1986, the USEPA established the current regulatory basis for fluoride in drinking water with a MCL of 4 mg/L and a SMCL of 2 mg/L. The MCLG and MCL for fluoride are both 4 mg/L. This concentration is the level below which a lifetime of exposure is not expected to cause health problems associated with skeletal fluorosis, a crippling disease, and is an enforceable maximum allowable concentration under the SDWA. As of the publication of this manual, USEPA was in the process of reviewing its criteria for fluoride.

The USEPA has established the secondary drinking water regulations with associated SMCLs, which are nonenforceable guidelines. The secondary contaminant list is based on aesthetic (taste, odor, staining of clothes or materials) or cosmetic effects (skin or tooth discoloration). Fluoride is the only contaminant other than copper that has both a primary and secondary standard. The SMCL for fluoride is 2 mg/L and is meant to prevent discoloration or pitting of teeth in children exposed during their formative years to water with high naturally occurring levels of fluoride. The SMCL seeks to balance the beneficial effects of fluoridation (protection against dental caries) with the undesirable cosmetic effects caused by excessive exposure. A state may choose to adopt a secondary standard (or any level lower than the federal MCL) as its enforceable standard.

The SDWA requires that water utilities notify customers in their annual Consumer Confidence Report when any regulated contaminant exceeds the regulatory limit in its treated water, regardless of whether the contaminant is naturally occurring or added as part of the treatment process. Any public water system that exceeds the primary or secondary MCL for fluoride should consult with its primary drinking water administrator to clarify the specific requirements for public notification. In general, an exceedance of the 4 mg/L MCL must meet the Tier 2 public notification requirements. An exceedance of the 2 mg/L SMCL must meet the Tier 3 public notification requirements. Because some states use the SMCL as the enforceable standard, the public notification requirements may differ from state to state. Any water utility serving a community considering fluoridation should have a firm understanding of the notification requirements in its state prior to initiating fluoridation.

State and Local

States have used different strategies to implement community water fluoridation programs, including legislation, administrative regulation, and public referenda at the state or local level. Thirteen states require fluoridation of public water systems that do not naturally contain optimal levels (Table 1-1). The specific requirements of these laws vary from state to state. Some have population thresholds, such that only systems serving a minimum number of customers are required to add fluoride, while others allow individual communities to opt out.

Ohio and Kentucky provide a good example of the variation between states. Ohio's state fluoridation statute requires all community water systems serving more than 10,000 customers to adjust fluoride to optimal levels. In contrast, Kentucky's regulation is an administrative code that imposes different requirements based on the population size of the community served.

Table 1-1 States requiring fluoridation of public water supplies

State	Year of mandate
Connecticut	1965
Kentucky	1966
Illinois	1967
Minnesota	1967
Ohio	1969
South Dakota	1969
Georgia	1973
Nebraska	1973, 2008
California	1995
Delaware	1998
Nevada	1999
Louisiana	2008
Arkansas	2011

Source: CDC, www.fluidlaw.org

In addition to the states listed in Table 1-1, Washington, D.C., and Puerto Rico have fluoridation mandates. The US Army Corps of Engineers Washington Aqueduct (the wholesale water supplier to the District of Columbia Water and Sewer Authority; Arlington County, Va.; and the City of Falls Church, Va.) has been adding fluoride since 1952.

Puerto Rico enacted legislation in 1998 requiring water fluoridation as needed to bring the fluoride concentration to an optimal level. The standards are governed by the Department of Health and operational compliance is overseen by the Aqueduct and Sewer Authority. Although Puerto Rico requires community water fluoridation, it is not widely practiced primarily due to infrastructure issues.

Although only 13 states have a fluoridation mandate, no state prohibits fluoridation. In most states, the decision is made at a local (regional or municipal) level. The decision at a local level can arise and be implemented in several ways. As an example, a local municipal health agency may make the recommendation to a city's governing body (i.e., mayor, city council, etc.). The local legislature may put the recommendation to referendum as a ballot question during a future election. Some municipalities have local ordinances requiring fluoridation. A local ordinance may be instituted if the legislature grants the recommending health agency the authority to require fluoridation without a referendum. The issue can become more complex when a public water supplier serves more than one municipality, such as in consecutive systems.

In most cases, the water supplier is not the entity that authorizes fluoridation, although it is tasked with implementation. The mandate to fluoridate can be the result of a legislative act or a directive from a state or local health agency. Regardless of the mechanism by which fluoridation was instituted, the water supplier is responsible for working with its primary regulatory drinking water administrator to ensure compliance with relevant permit requirements, such as design and maintenance of the fluoride chemical feed system, monitoring of fluoride levels in water samples, reporting data, and public notification.

The obligation a utility may have under fluoridation requirements is likely to be different in different communities. It is imperative that the water utility in any community considering fluoridation consult with its state drinking water administrator throughout the process of consideration and implementation of fluoridation.

Regulations in Other Countries

Similar to the SDWA in the US, the European Water Quality Directive (1980) provides maximum allowable concentrations of substances in water, and European countries use this directive as the framework for establishing their own water quality regulations. Currently, the WHO's guideline for fluoride is 1.5 mg/L. The directive neither mandates nor prohibits fluoridation. In the UK, the decision to fluoridate is made at a local level. The fluoridating utility works to a Code of Practice for fluoridation, which provides standards for the water industry. Compliance with the standards is monitored by the Drinking Water Inspectorate (the regulator for drinking water quality). Most recently, the European Parliament on May 16, 2006, voted 526 in favor and 126 against to maintain fluorides on the approved list of additives.

As in the United States, the fluoridation of public water supplies in Canada is not mandated nationally. Health Canada is the agency that oversees drinking water quality and works with the provincial and territorial governments to review, maintain, and periodically revise the Guidelines for Canadian Drinking Water Quality. Canada's maximum acceptable concentration (MAC) of fluoride in drinking water is 1.5 mg/L. As with organizations in the United States, Health Canada has recently revised its recommendation that the optimal level of fluoride in drinking water (for protection against dental caries) be decreased from a range (0.8 to 1.0 mg/L) to the single value of 0.7 mg/L. The decision to fluoridate is made by municipalities in collaboration with their provincial or territorial authority, and may include input from residents. Health Canada, the Canadian Public Health Association, the Canadian Dental Association, and the Canadian Medical Association all endorse the use of fluoride for the prevention of dental cavities. More information on water fluoridation in Canada can be found in *Fluoride and Human Health*, Health Canada, October 2010.

REFERENCES

Susan Griffin, K. Jones, and S.L. Tomar. 2001. An Economic Evaluation of Community Water Fluoridation. *Journal of Public Health Dentistry* 61(2):7886. http://onlinelibrary.wiley.com/doi/10.1111/j.1752-7325.2001.tb03370.x/pdf.

DeSimone, L.A. 2009. *Quality of Water from Domestic Wells in Principle Aquifers of the United States*. Reston, VA: US Geologic Survey.

Center for Disease Control. 2012. *Fluoridation Growth*. Atlanta, GA: CDC.

Health Canada. 2010. Fluoride and Human Health. Ottawa, ONT; Health Canada.

Chapter 2

Health and the Human Body

This chapter focuses on the health effects of fluoride and, in particular, the role of fluoride in preventing tooth decay.

FLUORIDE IN THE HUMAN BODY

The mineral and hard tissue structure of bones, teeth, and other parts of the skeleton contain fluoride. Soft body tissues do not retain fluoride, except as found in calcified deposits. Bones and teeth are predominantly composed of a crystalline matrix of calcium, magnesium, phosphorus, hydroxide, and fluoride. Since the fluoride ion and hydroxide have roughly the same atomic diameter, they can interchange on the mineral crystal. As the relative amount of fluoride increases and hydroxide decreases, bone density increases and tooth enamel has increasing resistance to decay. Fluoride acts on teeth both topically and systemically, although the topical mode of action is considered more important for reduction of dental caries. The topical effect can occur either through direct contact of fluoridated water with teeth or as a result of slightly elevated background fluoride concentrations in the saliva; the fluoride is then available to be incorporated at the tooth surface. The systemic effect occurs when fluoride diffuses through the walls of the stomach and intestines. The fluoride level in the blood is then available to all the skeletal structures, including the teeth, and in saliva. Fluoride not deposited in bones and teeth is excreted through the kidneys.

Fluoride incorporation by the teeth systemically occurs during the time of a child's formation and growth as the teeth are forming in the gum, pre-eruption. Teeth differ from other parts of the skeleton in that once they are formed, with the exception of the inner part of the tooth and the root, there is very little cellular activity. Maximum reduction in caries is achieved when fluoride is available both via the blood supply for incorporation during tooth formation and after the tooth erupts topically at the tooth surface over the lifetime of the individual.

The surface of the tooth is always exposed to a demineralization and remineralization cycle. Food and bacteria can result in a demineralization of the tooth surface, but fluoride in saliva and plaque promotes a remineralization of the tooth surface with a competent crystal. Although the systemic exposure allows the tooth enamel to form as a decay-resistant crystal, topical exposure allows a repair of tooth surface demineralization and therefore benefits all people of all ages.

The skeleton is the fluoride reservoir. When fluoride is ingested, roughly half of the fluoride is incorporated in the skeleton and roughly half is expelled principally by filtering in the kidneys, which quite efficiently removes fluoride from the blood. The bones experience a demineralization and remineralization throughout a person's lifetime. That process allows fluoride to be incorporated into the bone, but it also allows release of fluoride when ingested fluoride exposure reduces. This results in replenishment of fluoride in the bloodstream if fluoride exposure is less than it was at an earlier point. Although benefits of fluoride for the skeleton structure is not fully understood, the 2006 National Research Council (NRC) review of fluoride in drinking water cited some studies that suggest that the fluoride level for good oral health may be beneficial for bones in that there appears to be a relationship with lower bone breakage when the water is optimally fluoridated (National Research Council 2006).

DISCOVERY OF THE BENEFITS OF FLUORIDE

The modern study of the impacts of fluoride on the human body had its beginnings in the late 19th century with independent observations of mottled enamel by Stefano Chiaie in Italy and C. Kuhns in Mexico. As early as 1901, J.M. Eager documented his findings of discolored teeth in Naples, Italy, and postulated that the etiology may be connected to either something in the air or drinking water (although the source was believed to be volcanic fumes or the "emanations of subterranean fires"). Also in 1901, a young dentist named Frederick McKay relocated from Philadelphia to Colorado Springs, Colo. spent the next several decades researching mottled enamel, fluoridation, and the prevalence of dental caries.

Upon arrival in Colorado Springs, McKay noticed that many of his patients had significant brown staining, or mottling of the teeth, an enamel appearance he had not previously observed. This mottling was later termed *dental fluorosis*. A study by the Colorado Springs dental society completed at the time indicated that up to 90 percent of the city's locally born children had this condition, which McKay termed "Colorado brown stain." In 1908, McKay began conducting epidemiological studies into the cause of the staining in collaboration with G.V. Black, dean of the Northwestern University Dental School. They determined that the mottled enamel was caused by developmental imperfections in the children's teeth. McKay later concluded that these teeth were surprisingly resistant to decay.

After years of field investigations, McKay began to believe that the etiologic agent responsible for the mottled enamel was something in the water supply. In 1923, McKay recommended to the residents of Oakley, Idaho, that they change their water source due to the elevated levels of fluorosis observed in the town's children. Within a few years of the change, the younger children of Oakley were not exhibiting mottled enamel. A similar scenario occurred in Bauxite, Ark. In both cases, the water supply was analyzed but nothing unusual was observed. In 1931, McKay's theory was validated when he sent water samples to H.V. Churchill, a chemist with Aluminum Company of America. Churchill was using a new method of spectrographic analysis that could evaluate the concentration of a variety of analytes (including fluoride) down to much lower concentrations than the method previously employed by McKay. All of the samples McKay submitted were from areas where mottled enamel was endemic, and results of all samples showed elevated levels of fluoride (2 to 12 mg/L).

Once the agent causing fluorosis had been identified, research into the occurrence and dose-response relationships could commence. In the 1930s, H. Trendley Dean, the head of the Dental Hygiene Unit of the US Public Health Service (PHS), and McKay collaborated to better understand the adverse health effects of fluoride in drinking water. As the investigations continued with few adverse effects but increasing evidence of the benefits, their research continued to determine optimal fluoride concentrations in drinking water to balance the benefits of fluoride with the adverse effects. Dean identified locations around the United States with naturally occurring levels of fluoride, establishing a community fluorosis index. This led to further studies that quantified the dose-response relationship between fluoride, mottling, and the reduction of dental caries. The 21 Cities study in 1942 by Dean used environmental observations to establish a basis for an optimal level of fluoride for oral health benefits. From this research stemmed an optimal level of fluoride that both maximized caries reduction in a community while minimizing adverse health effects. Three distinct relationships were recognized:

1. At fluoride levels up to 1.5 mg/L, a reduction in tooth decay was observed. Fluoride levels greater than 1.5 mg/L did not significantly decrease the incidence of decayed, missing, or filled teeth. However, levels greater than 2.0 mg/L could increase the occurrence and severity of mottling.
2. At a fluoride level of approximately 1.0 mg/L, the optimal condition exists: maximum reduction in caries with no aesthetically significant mottling.
3. At fluoride levels below 1.0 mg/L, caries reduction diminished; at levels less than 0.5mg/L, approximately four times higher occurrence of caries was noted than in children exposed to levels greater than 0.5 mg/L.

Community water fluoridation studies in the 1940s were initiated to test the hypothesis that dental caries could be prevented by adjusting the fluoride level of community water supplies from negligible levels up to 1.0 to 1.2 mg/L. Studies were conducted in four pairs of cities in the United States and Canada, and caries were reduced 50 percent to 70 percent among children in communities with fluoridated water. These studies established fluoridation as a practical and effective public health measure.

In 1962, the USPHS set the recommended range of fluoride in drinking water from 0.7 mg/L (in warmer climates) to 1.2 mg/L (in colder climates). A range was recommended because average water consumption was believed to vary with average daily air temperature with an increased consumption of water in warmer locations. In January 2011, the US Department of Health and Human Services proposed as the recommended level of fluoride for community water systems a single value of 0.7 mg/L, which is the lowest level of the accepted beneficial range. The basis for setting the recommended value near the lowest recognized beneficial fluoride level is twofold: (1) Multiple sources of fluoride contribute to modern exposure levels, and (2) studies conducted since the recommended range was first introduced indicate that water consumption is not dependent on ambient air temperature as was once believed.

Dental Health

Dental caries, or tooth decay, is a largely preventable bacterial disease. Nearly everyone suffers from dental caries, making tooth decay the most prevalent chronic human disease. Before the widespread use of fluoride, more than 98 out of every 100 Americans experienced tooth decay by the time they reached adulthood. Tooth decay begins in early childhood and reaches a peak in adolescence, with the highest incidence of tooth decay being found in school-aged children. In the United States, tooth decay is experienced by more than 40 percent in the primary teeth of children ages two to 11 years, more than 60 percent in the permanent teeth of adolescents 12 to 19 years, and 91 percent of adults 20 to 64 (Dye et al. 2007). In the age group of 12- to 19-year-olds, the average US resident will have 2.55 decayed, missing, or filled teeth (Dye et al. 2007).

Tooth decay is a function of the presence of sugar as a substrate and certain bacteria combined with dental hygiene and fluoride. Cavities, or caries, form when acids dissolve the tooth enamel, thus destroying the tooth structure. These acids are produced as a metabolic by-product of certain bacteria, in particular *Strepotococcus mutans* and *Lactobacilli*. Cariogenic bacteria (i.e., bacteria that cause dental caries) reside in dental plaque, an organic matrix of bacteria, food debris, and salivary components that adhere to tooth enamel. Although plaque serves as a substrate for cariogenic bacteria, it is also the mouth's fluoride reservoir. As food is ingested, the bacteria in plaque consume the sugars, and this produces organic acids that lower the pH in the mouth. When this occurs, plaque releases fluoride onto the tooth surface. The presence of fluoride augments the remineralization process, which is discussed in greater detail in the subsequent paragraphs.

A cycle of demineralization and remineralization is constantly taking place in the mouth (Figure 2-1). Demineralization involves the loss of calcium, phosphate, and carbonate from the tooth's enamel structure; if it is not halted or reversed, it will eventually result in the formation of a cavity. Remineralization is the reversal of this process; it occurs when calcium and phosphate in the saliva are deposited back into the enamel. Fluoride enhances remineralization by adsorbing to the tooth surface and attracting both calcium and phosphate ions. Fluoride also acts to bring the calcium and phosphate ions together and is included in the chemical reaction that takes place, producing a crystal surface that is much less soluble in acid and more resistant to decay than the original tooth material. Demineralization and remineralization are constantly taking place on the surface of the teeth throughout a person's life.

Fluoride from topical sources also inhibits the cariogenic bacteria themselves. Cariogenic bacteria take up fluoride when they produce acid, and once inside the cells, fluoride interferes with bacterial enzyme activity and ultimately their intracellular pH control. This reduces bacterial metabolism and acid production, which directly reduces the dissolution rate of tooth mineral.

Adverse Health Effects

The USEPA's MCL for fluoride (4 mg/L) is meant to be protective of human health and is much higher than the dose considered optimal for prevention of dental caries (0.7 mg/L). The weight of evidence from all currently available research does not support a link between exposure to fluoride in drinking water at 0.7 mg/L and any significant adverse health effects, including those related to cancer, immunotoxicity, reproductive/developmental toxicity, genotoxicity, and/or neurotoxicity (National Research Council 2006). However, elevated fluoride concentrations have been linked to adverse health effects, some of which are discussed further in the following sections.

Dental Fluorosis. Dental fluorosis is caused by a hypo-mineralization in enamel formation. It is a permanent condition that results from exposure to increased levels of fluoride during tooth development. Because dental fluorosis is a condition that occurs when teeth are forming pre-eruption in the gums, only children age eight years old or younger are at risk; children older than eight years, adolescents, and adults are not susceptible to dental fluorosis.

Dental fluorosis becomes apparent upon eruption of the teeth. It ranges from very mild symmetrical whitish areas on teeth (very mild dental fluorosis) to pitting of the enamel and brownish discoloration (severe dental fluorosis). The very mild and mild forms are barely detectable even by experienced dental professionals and have no effect on tooth function. Moderate and severe forms of dental fluorosis, which the USEPA has

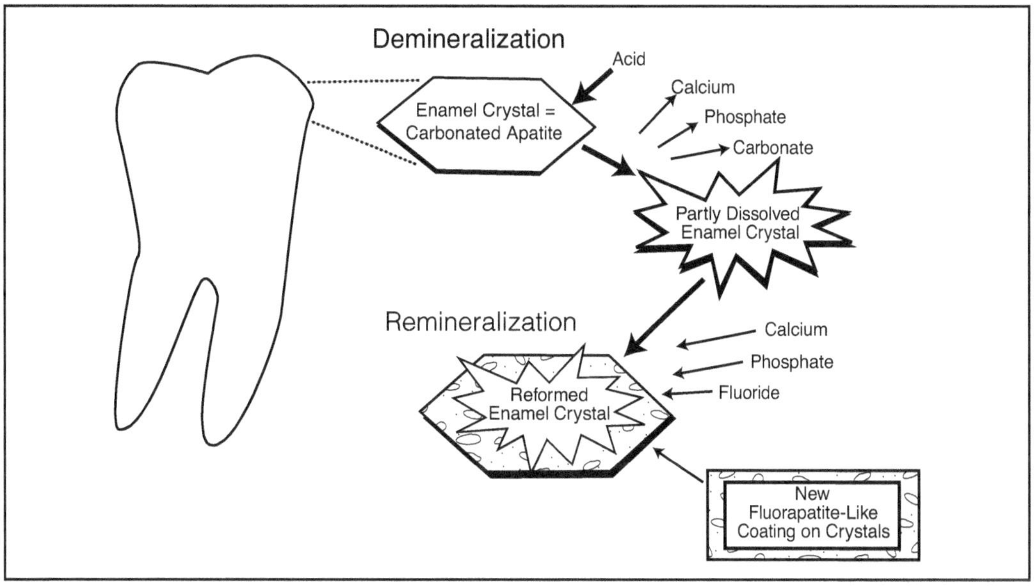

Figure 2-1 The demineralization and remineralization process

Source: Adapted from Featherstone, J.D.B. 1999.

considered a cosmetic issue, represent less than 3.5 percent of the cases of fluorosis nationally based on the National Health and Nutrition Examination Survey (Beltran-Aguilar et al. 2010). Figure 2-2 illustrates that a small percentage of the population is affected by very mild and mild fluorosis at 1 mg/L fluoride, which corresponds to the concentration at which the optimal benefit from dental caries reduction is achieved. Recent studies have shown a slight increase in dental fluorosis in communities with nonfluoridated water. This may be the result of other sources of fluoride, including toothpastes, dietary supplements, mouth rinses, professional applications, and foods and beverages processed with fluoridated water.

Young children and dental fluorosis. The principal cause of moderate and severe dental fluorosis is from children swallowing fluoride toothpaste. For children who use toothpaste, there is an increased chance of dental fluorosis among those younger than six years, and especially for those younger than two years, because they are more likely to swallow the toothpaste than older children as their spitting reflex has not yet developed. To prevent fluorosis, the CDC recommends that health-care professionals provide direct individual advice on the use of fluoride toothpaste by young children, especially those younger than two years, depending on individual assessment of diet, availability of dental health care, fluoride levels in the community water, and other risk factors for dental decay. Universally, dental health professionals should advise parents to supervise children while brushing teeth and to use only a smear or pea-size quantity of fluoride-containing toothpaste until age six.

Skeletal fluorosis and bone fractures. Fluoride is readily incorporated into the crystalline structure of bone and will accumulate over time. A small exposure of fluoride, as associated with water fluoridation, is generally considered beneficial for bones, but excessive exposure at levels above the recommended level for good oral health can result in adverse effects. Skeletal fluorosis is an adverse health effect caused by an accumulation of excessive fluoride in bone. This is the result of excessive chronic fluoride exposure.

Skeletal fluorosis is extremely rare in the United States with only seven reported cases by the Centers for Disease Control (CDC). However, fluorosis does occur in countries with areas of very high levels of naturally occurring fluoride, such as India and China.*

* Other confounding factors, such as inhalation of fluoride fumes from burning coal indoors, are often also present, thus increasing exposure levels.

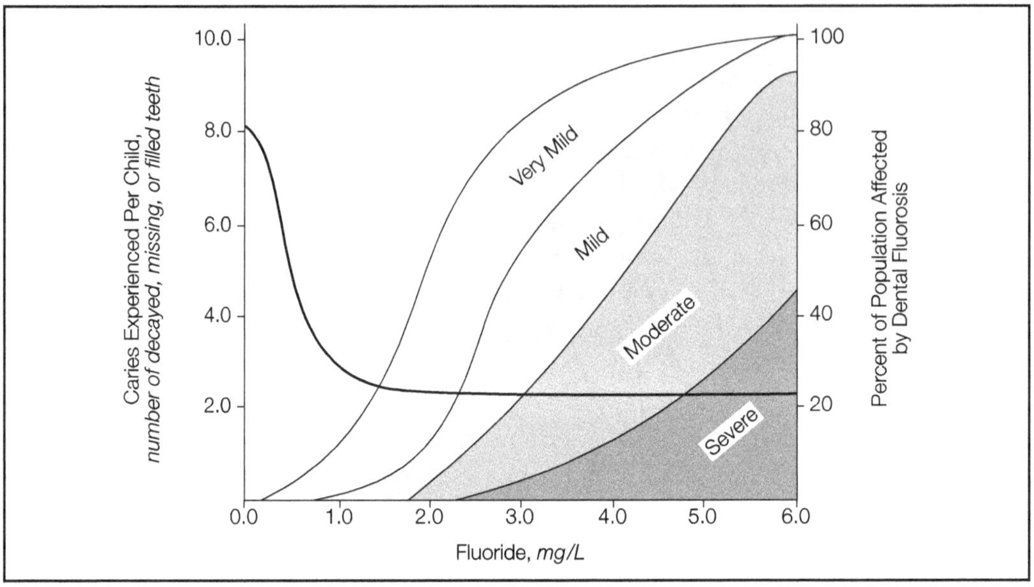

Figure 2-2 Dental caries and dental fluorosis in relation to fluoride in public water supplies

Source: Water Fluoridation: A Manual for Engineers and Technicians *(US Department of Health and Human Services, Centers for Disease Control and Prevention, September 1986).*

The Institute of Medicine (IOM) established recommendations for the tolerable upper intake level for fluoride to safeguard against skeletal fluorosis. This value is 10 mg/day for 10 or more years. This corresponds to a drinking water fluoride concentration of 5 mg/L, assuming a 2-L/day rate of water consumption. Fluoride increases bone density but can increase brittleness in long slender bones. Fluoride has been shown to affect bone calcification, and some studies have linked chronic elevated doses to bone fractures.

Cancer

Numerous human epidemiological studies on the potential carcinogenic effects of fluoride in drinking water have been conducted. The general consensus among the reviews completed to date is that no evidence exists linking water fluoridation and cancer.

According to a 2006 NRC review of fluoride in drinking water, the studies that had been completed to date regarding fluoride exposure and osteosarcoma risk were inconclusive and tentative. The NRC study went on to state that the multiyear Harvard-based hospital study that was in progress at the time would be an important assessment of the potential relationship between osteosarcoma and fluoride. The results from the full study did not find a significant association between bone fluoride levels and osteosarcoma risk (Kim et al. 2011).

In 2011, California's Office of Environmental Health Hazard Assessment published a report on the carcinogenicity of fluoride and its salts. This report concluded that the current body of epidemiological evidence on the carcinogenicity of fluoride is inconclusive, suggesting that fluoride is not a recognized causative agent for cancer.

Other Health Effects

Myriad other possible health effects related to fluoride have been studied, including its potential impact on kidney disease, hypersensitivity (allergic reactions), gastrointestinal problems, intelligence, thyroid (goiter) effects, effects on bioavailability or toxicity of toxic metals (aluminum and lead), enzyme and mutagenic (gene-altering) effects, and birth defects. Overall, evidence suggests no risk to people consuming optimally fluoridated water in the United States.

FLUORIDE DELIVERY

Recommended Water Fluoride Content

For water utilities, the recommended fluoride level is usually established by the appropriate state regulatory or health agency. The US Department of Health and Human Services (USHHS) recommended level of 0.7 mg/L can be used as a guideline if state limits have not been established. The 0.7 mg/L level was based on the evidence that in our modern society, water consumption does not significantly vary by region or climate, and that the lowest level in the recognized optimum range was sufficient protection for most persons.

Alternative Methods of Fluoride Delivery

While community water fluoridation is one of the most cost-effective mechanisms for dental caries reduction, other sources of fluoride also reduce tooth decay. At an individual level, people may receive fluoride for tooth decay prevention through fluoridated toothpaste and mouth rinses, professional fluoride treatments, fluoride dietary supplements, and beverages and food processed in fluoridated areas. Water fluoridation continues to be effective in reducing dental decay by 20 percent to 40 percent despite the availability and use of these other fluoride sources. Community fluoridated water also reduces the disparity in dental decay that may be seen among populations with low socioeconomic status, especially those with limited access to dentifrices, supplements, and routine dental care. Other forms of community fluoridation include salt and milk fluoridation.

REFERENCES

Beltrán-Aguilar E.D., L. Barker, B.A. Dye. 2010. Prevalence and Severity of Dental Fluorosis in the United States, 1999–2004. NCHS data brief 53. Hyattsville, MD: National Center for Health Statistics.

Dye, B.A., S. Tan, V. Smith, B.G. Lewis, L.K. Barker, G. Thornton-Evans, et al. 2007. Trends in Oral Health Status: United States, 1988–1994 and 1999–2004. National Center for Health Statistics. Vital Health Stat 11(248).

Featherstone, J.D.B. 1999. Prevention and reversal of dental caries: Role of low-level fluoride. *Community Dent Oral Epidemiol* 27:31–40.

Kim, H.Y., G. Chung, H.J. Jo, Y.S. Kim, Y.C. Bae, S.J. Jung, J-S Kim, S. Bo. Oh. 2011. Characterization of Dental Nociceptive Neruons. *Journ. of Dental Research* 90(6): 771–776.

National Research Council, U.S. Committee on Fluoride in Drinking Water, National Academies. 2006. Fluoride in Drinking Water: A Scientific Review of EPA's Standards.

US Department of Health and Human Services, Centers for Disease Control and Prevention. September 1996. Water Fluoridation: A Manual for Engineers and Technicians.

This page intentionally blank.

Chapter 3

Fluoride Products

This chapter provides an overview of the fluoride products used to adjust fluoride content in drinking water, presents their characteristics and sources, and introduces standards and regulations to ensure quality.

CHEMICAL CHARACTERISTICS

Historically, several products were used for water fluoridation during the first decades of water fluoridation. As water fluoridation practice evolved, now only three products are approved for adjusting fluoride:

- Sodium fluoride: a granular salt dissolved into a solution
- Sodium fluorosilicate: a granular salt dissolved into a solution
- Fluorosilicic acid: a liquid acid normally added without dilution

Over time, the trend has been a shift from dry products mixed into solutions to the predominant use of fluorosilicic acid. This trend for fluoride products mirrors the change in other additive products used for water treatment because the water treatment industry often has been willing to pay more for a premixed additive solution to minimize the handling of chemicals at a water treatment facility. The CDC has reported trends in use of products (Table 3-1).

This section provides background information about the manufacturing processes, aquatic chemistry dissociation of the fluoride ionic form(s), and physical and chemical characteristics.

Sodium Fluoride

The first fluoride compound used to adjust fluoride in drinking water was sodium fluoride. It was selected because (1) it is a readily available product and (2) its toxicity and physiological effects had been studied. Once fluoridation became an established practice, other compounds came into use.

Table 3-1 Trends in the use of fluoride products over time

	Population receiving adjusted fluoridated water	Percent of population receiving adjusted fluoridated water	Systems	Percent of systems
1960				
Fluorosilicic acid	10,285,630	25%		
Sodium fluorosilicate	24,588,937	60%		
Sodium fluoride	5,186,351	13%		
Ammonium silicate calcium fluoride	73,000	0.2%		
1993				
Fluorosilicic acid	80,019,175	63%	5,876	59%
Sodium fluorosilicate	36,084,896	28%	1,635	16%
Sodium fluoride	11,701,979	9%	2,491	25%
2010				
Fluorosilicic acid	152,501,133	81%	9,125	75%
Sodium fluorosilicate	24,224,615	13%	1,208	10%
Sodium fluoride	12,815,339	7%	1,825	15%

Source:

1. 1960 US Public Health Service Fluoridation Census
2. 1993 CDC Fluoridation Census
3. 2010 CDC Water Fluoridation Reporting System

Sodium fluoride is a white, odorless product available either as a powder or as crystals of various sizes. It is most commonly produced by neutralizing fluorosilicic acid with caustic soda (NaOH). It is also produced by neutralizing hydrogen fluoride with caustic soda. It is the most expensive of the fluoride additives because of the high processing effort and large amount of caustic soda required to produce it. As a result of the high cost of production, production sources are currently located outside the United States (CDC 2012).

Crystalline sodium fluoride is produced in various size ranges, usually designated as coarse or fine. The crystalline type is preferred for utilities practicing manual handling because it minimizes dusting when emptying a bag into a saturator. The AWWA product specification on sieve analysis of sodium fluoride specifies passing 98 percent through a US Standard Sieve Series mesh No. 20 and 50 percent through a No. 100 mesh sieve. Different manufacturers have different gradations, so some consumers request limits on allowable fines. Customer-specified limitations have often been to limit fines of less than 5 percent passing through a No. 325 mesh and less than 20 percent passing through a No. 100 mesh to minimize dusting of the product during handling. Sodium fluoride is commercially available in both 50- and 100-lb (23- and 45-kg) bags.

Sodium fluoride must be made into a solution before being added to water supplies. The chemical can be dissolved in a mixing tank, a saturator, or a dissolving chamber. (Equipment requirements are discussed in detail in chapter 5.) Its formula weight is 42.0 (fluoride is 19 and sodium is 23) and results in an available fluoride content of 45 percent. Its solubility is practically constant at near 4 g/100 mL of water at temperatures generally encountered in water supplies. It is the ideal additive for use with saturators for small systems because of its relatively constant solubility. It also dissolves readily and can achieve a saturated solution within five minutes. When added to water, sodium fluoride dissociates into sodium and fluoride ions:

$$NaF \Leftrightarrow Na^{+} + F^{-} \tag{3-1}$$

Solution pH varies with the type and amount of impurities, but solutions prepared from the usual grades of sodium fluoride exhibit a pH near neutral (approximately 7.6). Sodium fluoride is available in purities ranging from 90 percent to more than 98 percent. Impurities consist of water, free acid or alkali, sodium fluorosilicate, sulfites, and iron, plus traces of other substances.

Fluorosilicic Acid

Fluorosilicic acid is principally collected as a by-product of phosphate fertilizer. A small quantity, less than 5 percent of the market availability, is derived from gas separation columns from the production of hydrogen fluoride derived from calcium fluoride ore, and less than 1 percent of the market is derived from quartz etching of silicon chips. Fluorosilicic acid produced from calcium phosphate ore and apatite ores is created by adding the ores to sulfuric acid to form a phosphoric acid–gypsum slurry. The phosphoric acid is separated from the gypsum and desiccated to pellets for use in fertilizers. The fluoride remains dissolved in the phosphoric acid–gypsum slurry as a silica tetrafluoride gas. It can be removed from the slurry using vacuum extraction, then partitioned from the phosphate in gas separation columns with collection in a water bath. The collected acid is then concentrated to between 20 percent and 30 percent strength, with 23 percent being the reference concentration.

Acid prepared from phosphate rock (as part of fertilizer manufacture) contains colloidal silica in varying amounts. Although this is of little consequence when the acid is used as received, dilution results in the formation of a visible precipitate of the silica. Rather than attempt to adjust the acid strength with water dilution to some uniform figure, producers sell the acid as it is produced and adjust the price to compensate for acid strength greater or less than the quoted figure. Suppliers usually provide assay reports of the acid strength of each lot.

With a reference concentration of 23 percent, fluorosilicic acid contains a high proportion of water (typically 72 percent to 78 percent water content), and shipping large quantities can be expensive. It is normally collected during manufacturing in 20,000-gallon railcars that are then used to transport the product to regional depots for distribution to users. Large users can purchase the acid directly from the manufacturer in bulk lots (tank cars or tank trucks), but small users must obtain the acid from distributors, which usually package it in drums or polyethylene carboys. Some locations may have a regional distributor that uses a mini-bulk delivery of chemicals. Mini-bulk uses 500-gallon tanks on a truck that can then act as a nurse-truck delivery to top off on-site storage tanks of smaller facilities.

Fluorosilicic acid is not a pure substance but a mixture of various oligimers and crown-ether silico-fluoride complexes with an average formula of H_2SiF_6 and a formula weight of 144.0. This results in an available fluoride ion content of 79 percent. The most common concentration used in drinking water is 23 percent to 25 percent but the concentration may be higher or lower as delivered. Fluorosilicic acid is a transparent, fuming, corrosive liquid having a pungent odor and an irritating action on the skin; but based on impurities, it can vary from a water-white to a straw-colored appearance. On vaporizing, the acid decomposes to form hydrofluoric acid (HF) and silicon tetrafluoride (SiF_4). Under conditions where equilibrium between the fluorosilicic acid and its decomposition products exists, such as at the surface of strong solutions, etching of glass can occur as a result of the volatile hydrogen fluoride. At normal room temperature, hydrogen fluoride will volatilize from the surface of the solution. Due to the small percentage of fluoride in the form of hydrogen fluoride and the relatively small exposure of the surface to the bulk contents of a storage tank, hydrogen fluoride in fluorosilicic acid has a slow release rate. However, the corrosive character of hydrogen fluoride results in a nuisance etching and corrosion to glass, metals, and electrical products if tanks are not property sealed and vented.

When fluorosilicic acid is dissolved in water, the dissociation of hydrofluoric acid and silicon tetrafluoride is pH dependent. At the pH and concentration used for water fluoridation, the presumed dissociation results in hydrofluoric acid and silicon tetrafluoride:

$$H_2SiF_6 \leftrightarrow 2HF + SiF_4 \quad (3\text{-}2)$$

Hydrofluoric acid strongly dissociates, lowering the pH of the solution:

$$HF \leftrightarrow H^+ + F^- \quad (3\text{-}3)$$

Silicon tetrafluoride is a gas that volatilizes if present in high concentrations. It also reacts quickly with water to form silicic acid (H_2SiO_3) or silicon dioxide (SiO_2):

$$SiF_4 + 3H_2O \leftrightarrow 4HF + H_2SiO_3 \quad (3\text{-}4)$$

$$SiF_4 + 2H_2O \leftrightarrow 4HF + SiO_2 \quad (3\text{-}5)$$

Concentrated solutions of fluorosilicic acid exhibit a low pH (approximately 1.2). The alkalinity consumption as mg/L of calcium carbonate ($CaCO_3$) per mg/L of fluoride is 2.08, so the buffering capacity of the water should be considered when adding fluoride to poorly buffered waters.

Attempts to dilute the acid are subject to errors in measuring both the acid and the diluting water. The acid should be used undiluted as it comes from the shipping containers. The most unstable range for dilution is a mix of 10:1 to 20:1 when the silica can readily precipitate into a nuisance scale.

Fluorosilicic acid has a higher density than water, and one of the two approved AWWA verification tests is hydrogen titration or gravimetric measurement of density to determine the concentration of fluorosilicic acid. Hydrogen titration is the more accurate method as it differentiates other acids from fluorosilicic acid in solution. Fluorosilicic acid can be delivered at a concentration between 20 percent and 30 percent, but is invoiced at a 23 percent basis, so it is important for an operator to measure the concentration of the solution in order to correctly establish the precise dosage. At a concentration of 25 percent, the density of the solution is 10.1 pounds per gallon (Figure 3-1).

Sodium Fluorosilicate

Fluorosilicic acid can readily be converted into various salts, one of which, sodium fluorosilicate, is a partially neutralized form of the acid. The most common process is to add sodium chloride to fluorosilicic acid, resulting in a hydrochloric acid–sodium fluorosilicate solution. Sodium fluorosilicate can also be formed by mixing fluorosilicic acid with sodium chloride or caustic soda. The sodium fluorosilicate is precipitated into crystals and then desiccated. In some cases, sodium fluorosilicate tends to be the least expensive product in use because the transport is a major factor in the cost of any additive product. Because sodium fluorosilicate does not have the large dilution water associated with fluorosilicic acid, it costs less to transport. The conversion of aqueous fluorosilicic acid to a dry product containing a high percentage of available fluoride results in a compound having

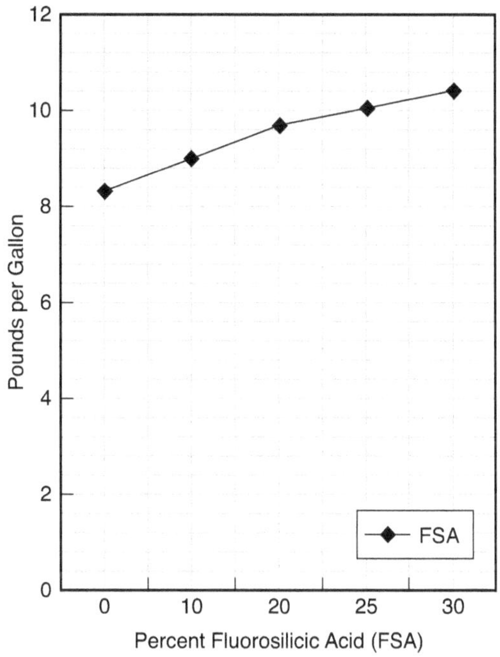

Figure 3-1 Density of fluorosilicic acid vs. percentage based on data provided by The Mosaic Company, 2006.

Source: CDC

most of the advantages of the acid with few of its disadvantages. It is also a weak acid in comparison to fluorosilicic acid, which is a strong acid.

The sodium fluorosilicate used in the United States is produced domestically and imported. Sodium fluorosilicate is normally packed in bags and drums similar to those used for sodium fluoride, but is also available in 2,000-pound supersacks for ease of handling. As with sodium fluoride, sodium fluorosilicate must be made into a solution in a dissolving chamber before being fed into the water supply.

Sodium fluorosilicate is sold with a density of about 85 lb/ft^3 (1,400 kg/m^3). A typical sieve analysis of the regular grade allows more than 98 percent through a US Standard Sieve mesh No. 40 and at least 25 percent through a 325-mesh sieve. Depending on feeding characteristics, other size specifications can be selected. Experience shows that a low moisture content plus a relatively narrow size distribution results in a material that is handled better by dry feeders.

Sodium fluorosilicate is a white, odorless, crystalline powder. Its molecular weight is 188 (two sodium at 46, six fluoride at 114, and one silica at 28) resulting in an available fluoride ion content of 60 percent. Its solubility is temperature dependent and varies from 0.44 g/100 mL of water at 32°F (0°C) to 2.45 g/100 mL of water at 212°F (100°C). When sodium fluorosilicate is dissolved in water, virtually 100 percent dissociation occurs within several minutes:

$$Na_2SiF_6 \leftrightarrow 2Na^+ + SiF_6^- \qquad (3\text{-}6)$$

Silicofluoride ions, SiF_6^-, may react in two ways. The more common reaction is hydrolysis of SiF_6^-, releasing fluoride ions and silica.

Table 3-2 Characteristics of fluoride products

Item	Sodium fluoride (NaF)	Sodium fluorosilicate (Na_2SiF_6)	Fluorosilicic acid (H_2SiF_6)
Form	Dry crystals	Dry crystals	Aqueous liquid
Molecular weight	42.00	188.05	144.08
Pounds required per million gallons (grams required per cubic meter) for 0.7 mg/L F^- at indicated purity	11.5 (14.2) (98%)	11.7 (1.07) (98.5%)	29.3 (3.5) (23%)
Sodium ion contributed at 0.7 mg/:L F^-	0.82	0.28	0.0
F^- ion storage space, ft^3/100 pounds (m^3/kg)	22–33 (0.0137–0.0212)	23–30 (0.0143–0.0187)	54–73 0.0337–0.0455)
Solubility at 25°C, g/100 mL water	4.05	0.568 at 10°C 0.632 at 15°C 0.696 at 20°C 0.762 at 25°C	Greater than 40,000
pH of saturated solution	7.6	3.5	1.2 (23% solution)
Alkalinity consumed, mg as $CaCO_3$ per mg F^-	0.0	1.06	2.08
Weight lb/ft^3	65–90 (1,000–1,400)	85–90 (1,320–1,400)	10.1 lb/gal at 25% 10.5 lb/gal at 30% (1.26 kg/L at 25%) (1.31 kg/L at 30%)
Shipping containers	50-lb bags 25-kg bags 125- to 400-lb fiber drums	50-lb bags 25-kg bags 124- to 140-lb fiber drums	20,000-gal railcar 4,000- to 5,000-gal truck 500-gal mini-bulk deliveries 300-gal tote tanks 55-gal drums 13-gal carboys
Typical commercial purity, percent	97–99	98–99	20–30

Source: ANSI/AWWA B701, B702, B703; AWWA Fluoride Standards Committee

$$SiF_6^- + 2H_2O \leftrightarrow 4H^+ + 6F^- + SiO_2 \qquad (3\text{-}7)$$

Alternatively, SiF_6^- dissociates very slowly, releasing fluoride ions and silicon tetrafluoride:

$$SiF_6^- \leftrightarrow 2F^- + SiF_4 \qquad (3\text{-}8)$$

Silicon tetrafluoride reacts as described in Eq 3-4 and Eq 3-5. The pH values of the solutions are acidic (generally about 3.6). Sodium fluorosilicate is available in purities of 98 percent or greater, with the principal impurities being water, chlorides, silica, and carbonates. The alkalinity consumption as mg/L of $CaCO_3$ per mg/L of sodium fluorosilicate is 1.06, so the buffering capacity of the water should be considered when adding fluoride to poorly buffered waters.

Table 3-2 presents the characteristics of fluoride products.

Product Standards

Two entities publish standards that are generally accepted by the water treatment industry: AWWA and the National Sanitation Foundation (NSF) International. Both publish in conjunction with the American National Standards Institute (ANSI). AWWA prepares standards for the manufacturing, quality, and verification of the fluoride additives for use in drinking water. The AWWA standards set the physical, chemical, and impurities standards for each product, including information on verification of the standard requirements and requirements for delivery. AWWA publishes the following standards for individual fluoridation products:

- ANSI/AWWA B701 Sodium Fluoride
- ANSI/AWWA B702 Sodium Fluorosilicate
- ANSI/AWWA B703 Fluorosilicic Acid

NSF International prepares standards that cover impurities of drinking water treatment additives from their production and through their distribution to user and includes documentation of the purity of additives. Standard NSF/ANSI 60 Drinking Water Treatment Chemicals – Health Effects provides for product purity and safety assurance that aims to prevent adding harmful levels of contaminants from chemicals and water treatment additives. Standard 60 sets the requirements for certification of individual drinking water treatment chemicals for compliance with MCL concentrations set by the USEPA; it also sets standards for chemical handling to ensure product quality is maintained. Verification testing by independent certification entities including NSF International and Underwriters Laboratories (UL), documents that the actual purity exceeds the standards. The NSF/ANSI Standards 60 and 61 (Drinking Water System Components – Health Effects) were developed by a consortium of associations, including NSF International, AWWA, ANSI, the Association of State Drinking Water Administrators (ASDWA), and the Conference of State Health and Environmental Managers to replace the previous additives overview effort by the USEPA.

In the United States, authority to regulate products for use in drinking water rests with individual states. NSF, in cooperation with the ASDWA, conducts a biennial survey to determine which states and Canadian provinces/territories require (by legislation, regulations, or policies) that drinking water additives be evaluated by NSF/ANSI 60. In 2013, an ASDWA survey found that 47 states and nine provinces/territories had adopted NSF/ANSI 60.

NSF/ANSI Standard 60 specifies the product quality for use in drinking water treatment, and various independent certification entities provide the validation for the products. Those third-party entities certify products in accordance with NSF/ANSI Standard 223. Standard 223 covers minimum requirements to be used by conformity assessment organizations when testing and certifying products to Standard 60.

Other Standards

US Pharmacopeia (USP) prepares standards for products used in pharmaceuticals. Some water facility operators may hear from concerned citizens that USP would provide a more rigorous quality standard. In fact, USP grade sodium fluoride provides less protection for drinking water because it is tested for nonspecific heavy metals, and it has no criteria for arsenic content or radiological exposure. Most importantly, USP does not provide certification of quality by an independent third-party credentialing entity to established quality verification. USP also does not have a standard for fluorosilicic acid or sodium fluorosilicate.

REFERENCE

Center for Disease Control. 2012. Water Fluoridation Principles and Practices. Atlanta, GA: CDC.

Chapter 4

System Planning

This chapter introduces the types of fluoridation systems. It also presents guidelines for system planning, including chemical selection, preliminary planning, dosage calculations, and system considerations.

Fluoridation systems, as is the case with other drinking water processes, are normally planned in coordination with the state or provincial agencies responsible for drinking water regulations. The regulations may address fluoride levels, recommended systems, monitoring requirements, and record keeping. The agencies may have the final approval of the system and design.

Among considerations in choosing the right fluoridation system for a particular situation are:

- Type of fluoridation system
- Water system characteristics (such as source of supply, single or multiple points of supply, treatment, storage, distribution, personnel)
- Chemical availability
- Cost

DECIDING WHETHER TO FLUORIDATE

The decision to fluoridate may be established on the state level, regional level, or local level, depending on the state and community. Water fluoridation is managed by a state or provincial program, but each program has different requirements with respect to the requirements for fluoridation. Currently, 13 states have a mandatory fluoridation requirement, although there may be caveats such as population thresholds, the stipulation that a public water system will implement fluoridation if another entity provides the funding, or if directed by the state to implement.

The first step in deciding to fluoridate is working with the state on requirements for a particular water system, including new startups and terminations. The next level of consideration is potential regional implications. In some cases, a water system may require concurrence of all other water systems with a connection in order to agree to the decision on fluoridation, so a regional coordination may be necessary. The local water system might have a directive to fluoridate by a public health board, by a local community governance board (if different than the water utility), or by a public vote.

If there are no state requirements for fluoridation, then it becomes a community decision. The question is strictly related to the health of the community. Communities with fluoridated water have significantly improved oral health than communities without fluoridated water, and that benefit extends to all people of all ages and socioeconomic demographics.

Choice of System

There are three basic types of fluoridation systems:

- Using saturated solutions of sodium fluoride
- Using unsaturated solutions of sodium fluorosilicate
- Feeding aqueous fluorosillic acid

Using Sodium Fluoride. Sodium fluoride is normally fed as a saturated solution. A saturator is a tank that allows the salt to dissolve to an aqueous saturated solution. Using a dry feeder instead of a saturator is uncommon based on reported systems by states to the CDC Water Fluoridation Reporting System, and this may be due to the higher relative cost of sodium fluoride and the suitability of sodium fluorosilicate in dry feeders.

With a nearly uniform saturation concentration across those temperatures at which most water is treated, sodium fluoride and the associated sodium fluoride saturator require minimal operator input to yield a consistent concentration of fluoride from the saturator. The process involves dissolving the salt into a saturated solution and then feeding that solution at a measured rate to the facility product flow. The cost for the equipment and installation is the least expensive configuration, it is easy to maintain, and operation is simpler than other systems. The downsides include the cost of the sodium fluoride, which is more expensive than the other fluoride products due to the greater processing to derive the product, and its limitations when using a high-hardness feedwater. The use of sodium fluoride in locations with high natural water hardness is a limitation as the high calcium in the water can suppress the saturated solution and result in increased nuisance scaling. For many small water systems, a saturator system may be an appropriate choice.

Using Sodium Fluorosilicate. Sodium fluorosilicate is a crystalline salt that must be dissolved to an unsaturated solution prior to feeding. An unsaturated solution is necessary because the salt achieves a saturated solution only after a lengthy mixing period. It also has a temperature-dependent concentration so the concentration of the unsaturated solution is indeterminate. Consequently, it cannot be used in a saturator but must instead be metered by a dry feeder into a solution tank where the unsaturated solution is formed. It is a weak acid, so there are fewer negative effects such as material attack on concrete metals, glass, and other surfaces. It has fewer personnel safety concerns and is easy to handle.

Sodium fluorosilicate can be the least expensive of the products depending on the cost of energy (fuel) and location because the logistical transport cost can be a significant portion of the cost of fluoridation products. As a weak acid, it can be more suitable in locations with high water hardness as it is not subject to the scaling and reduced solubility of sodium fluoride. Both sodium fluorosilicate and sodium fluoride have lower quantities of contaminants due to processing of fluorosilicic acid to derive the dry products, which tends to retain impurities in the reject slurry.

Feeding Aqueous Fluorosilicic Acid. Fluorosilicic acid is the most commonly used product. The handling characteristics of an aqueous product ready for direct addition can minimize operator involvement, and this is consistent with a general tendency in the water treatment industry to use liquid products not requiring mixing in water treatment. It is easy to simply have a storage tank that receives the delivered product and pump a metered acid feed to the application point. However, fluorosilicic acid is a strong acid and has an aggressive character with a propensity to attack many materials and surfaces. It also has the greatest potential for overfeeds and consequently requires more safety and regulators in the feed system.

FLUORIDE APPLICATION POINT

In systems that supplement naturally occurring fluoride, it may be necessary to fluoridate at each entry point to the distribution system. The configuration of the water system will be the major consideration on the type of feed arrangement that will be necessary.

If the water wholesaler does not already fluoridate, fluoridation application points may be required at all interconnections in regular use (i.e., not emergency connections). Once these categories of application points have been identified, the specific configuration of each application point can be determined.

The location of the fluoridation application point ahead of each entry point should be at a choke point, or bottleneck, through which all water entering the entry point will pass. In a water treatment plant, this location may be a channel where other water treatment chemicals are added, a pipe coming from the filters, or the clearwell. In a well–pump system, this may be the discharge line of the pump if there is only one. If there is more than one pump, the point may be the manifold line leading to the distribution system or storage facility.

When other chemicals are being fed, chemical compatibility must be considered. Although fluoride is nonreactive with most other chemicals used at a treatment facility, whenever possible, fluoride should be added after treatment in an effluent finishing location. Products containing aluminum, calcium, and phosphates can produce negative impacts when associated with fluoride. Chemical solutions containing calcium, including lime, have the potential to form a calcium fluoride scale. Therefore, the fluoride injection point should be as far downstream as possible from the addition point of those chemicals to allow adequate time for complete dissolution of the calcium before the fluoride is added. Facilities using aluminum sulfate (alum) can experience some fluoride loss if fluoride is added prior to the alum addition: at an alum dosage rate of 100 mg/L, fluoride loss can be up to 30 percent. Adding fluoride too early in the lime soda ash process can also result in a reduction in concentration. When possible, fluoride addition prior to phosphate addition for pipeline corrosion control is preferable for analytical verification because phosphate and other acids have a potential to interfere in the measurement of fluoride in some colorimetric testing procedures. If this situation cannot be avoided, then the facility should use the ion-specific electrode method of analytical measurement instead of the colorimetric method.

In most facilities, fluoride is added in a finishing step after other treatment processes. This is the preferred location so that nothing else will be added to affect the fluoride concentration. Less than 10 percent of facilities add fluoride earlier in the process flow, including the pretreatment process. In those cases, some chemical additions such as products containing calcium or aluminum have the potential to reduce the fluoride content. If this is the case, the facility should measure the fluoride in the water as it enters the distribution system to document the proper residuals.

Another factor in selecting the addition point is the desirability of having the feeder as close to the addition point as possible when dry fluoride is used and mixed into a solution. The discharge line from the feeder should be as short and straight as possible, although circumstances sometimes require a long discharge line. When this is the case, sharp curves or loops in the line should be avoided because they provide sites for precipitation buildup and subsequent blockage. Not infrequently, the addition of a separate unit process at an application point requires a new structure. The fluoride storage and feed equipment should have a designated room, and chemical storage should not be shared with different chemicals.

PROJECT PERMITTING AND PLANNING

Once the decision to proceed with water fluoridation is established, then the steps include identifying funding, preparing a design, obtaining all necessary permits, construction, and final implementation. Funding for implementation is often by the community or water system, but in some cases may be provided by another party.

Next, plans and specifications are prepared by the design engineer, which is usually a private professional design engineering firm retained by the water provider or, in a few cases, engineering staff of the water provider itself. In a few states, the state hires the engineer to provide design services for fluoridation systems to improve the consistency and quality of the installation. After approval by the permitting agency, the engineering plans and specifications become contract documents for a construction project that is bid and awarded to a qualified contractor. Some water providers may elect to purchase and install the fluoridation equipment using in-house staff and equipment.

Once the contract documents are in place, the fluoridated water system startup generally proceeds as follows:

- During construction, the design engineer often conducts field visits to observe construction procedures, answer contractor questions, and provide other engineering support to the contractor to ensure that the system is constructed according to the contract documents.
- As construction approaches completion, the water provider's operators receive training on fluoridation system equipment operation and maintenance from the design engineer and equipment vendors, or in some states, by state technical staff.
- The completed installation is inspected by the permitting agency for compliance with contract documents.
- The water provider prepares a fluoride monitoring and reporting plan for submittal to the permitting agency, and it notifies the public of the upcoming fluoridation implementation.
- The fluoridation system is ready to start up once the following happens: The permitting agency approves the installation and the monitoring and report plan, and the water provider completes a successful public notification.

Normal water fluoridation structures require a series of permits. Due to the presence of a hazardous chemical, these permits will be more extensive. The local building and safety department is likely to require:

- Building permits
- HVAC permits
- Plumbing permits
- Fire protection permits
- Electrical permits

In an effort to reduce the exposure of personnel to combustion products, the local fire department may require additional permits, including:

- Corrosive chemical storage facility permits
- Hazardous materials disclosure
- Sprinkler activation notification through a central data processing facility
- Tests of the activation notification

The successful construction of a fluoridation system is the result of efforts by and coordination among various entities, most notably regulatory and permitting agencies, the water provider, the fluoridation system design engineer, and the construction contractor.

Fluoridation implementation is heavily driven by regulatory requirements stipulated by permitting agencies, such as the state drinking water program and/or state health department, the local fire department, and others. The following permitting requirements are common in most jurisdictions that regulate fluoridation:

1. A permit from the state/regional/local regulatory agency is almost always required before the community can fluoridate.
2. Design drawings and specifications must be prepared (in many states by a licensed engineer) and submitted to the permitting agency for approval prior to construction.
3. Safety devices are required. These vary greatly among jurisdictions, but they generally include day tanks, overfeed controls, vacuum breakers, anti-siphon devices, and eyewashes and safety showers.
4. In many jurisdictions, a fluoridation system requires inspection and approval by the permitting agency's drinking water program staff before operation can commence. In some jurisdictions, the operators are also required to have some formal training prior to startup.
5. Most jurisdictions require personal safety gear for the safety of the facility operator, such as long gauntlet gloves, chemical aprons, respirators, goggles, showers, eyewash facilities.
6. All jurisdictions require the correct operator certification (appropriate for the size and complexity of the water system) prior to the operation of the fluoridation system.
7. Almost all jurisdictions require daily sampling for fluoride levels in the finished water. Many jurisdictions require daily sampling from each individual source (well and treatment plant). Water providers should understand the applicable sampling and reporting requirements. Some split sampling verification is generally required.
8. Water providers should check with the appropriate permitting agencies on requirements for records to be maintained, such as daily fluoride sample results at source, daily weights of fluoride chemical added, and, in some jurisdictions, calculation of fluoride levels.
9. Emergency response procedures should be specified for each facility.

PROCESS OF IMPLEMENTING FLUORIDATION

Designing a new system or evaluating an existing facility will require preparation of calculations. The following sections provide the basis for calculations for different types of facilities.

Process Calculations

Historically, deciding on the correct dosage for a particular location involved some consideration of ambient average temperature. Currently, that has been substantially simplified with the recommended level being established as a uniform 0.7 mg/L. Considering the natural fluoride in the source water, the dosage required to achieve a finished fluoride level of 0.7 mg/L can be calculated. The first step in planning is to determine the natural fluoride in the supply, and then that establishes the dosage required to achieve a finished fluoride level of 0.7 mg/L. Then, based on the system pumping rates, chemical purity, and available fluoride ions, the quantity of chemical needed daily can be determined.

Some treatment facilities will have a simple design, such as a groundwater well with a pump that either is on or off. Other facilities may have more complex and variable flows that accommodate different design considerations, including:

- Peak day flow
- Maximum month flow
- Annual average flow (this might be different based on current day flows at initial operations with a higher value at a future time when the facility is operating at design capacity with community growth)
- Minimum day flow
- Minimum hour flow (not common for water treatment, but could be a consideration in some cases)

An average flow rate determined from recent monthly data for three years should be used as a basis for current flows, along with the ultimate design capacity of the facility. Seasonal changes in water supply must be taken into account to determine the changes in water demand and average background water fluoride concentration. These inputs determine the fluoride dose, feed rates, chemical usage, and chemical storage. The CDC recommends that the water utility maintain a three-month supply on hand to guard against fluoride chemical shortages; the Canadian authority makes a similar recommendation.

Naturally occurring fluoride levels fluctuate for a variety of reasons, such as local geology, runoff, and so on. Therefore, the natural fluoride should be verified seasonally as it might change. For example, a wetter season (with rain dominating the flow) may result in a lower fluoride concentration, while a drier season (with groundwater springs dominating the flow) might result in a higher fluoride concentration. Most wells produce relatively uniform fluoride concentrations because the rate of fluoride leaching from the underlying geology tends to be consistent. Four to 10 data points per well should be sufficient, depending on the availability of data from the well. Many wells are not sampled more frequently than every three years, so older records may need to be accessed. When cluster wells are grouped, the combined fluoride concentration must be determined by a mass balance using the operating well flow rates under normal operational scenarios and conditions, not the maximum rated outputs.

Calculation of the chemical feed rate is necessary for proper adjustment of the fluoride feeder. To calculate the chemical feed rate, the following must be known:

- Desired fluoride dosage
- Plant rate or pumping rate
- Commercial purity and fluoride concentration of the fluoride chemical being fed (available fluoride ion)

The dosage is the amount of fluoride chemical needed to obtain the optimal fluoride level in the drinking water. The dosage, expressed as milligrams per liter (mg/L) or parts

Table 4-1 Typical values of purity and available fluoride ion (AFI)

Chemical purity	%	AFI	Available F %
Sodium fluoride	98	0.452	44
Sodium fluorosilicate	98.5	0.607	60
Fluorosilicic acid	23	0.792	18

per million (ppm), is obtained by subtracting the naturally occurring fluoride level from the desired fluoride level. For example, if the desired fluoride level is 0.7 mg/L and the natural fluoride level is 0.2 mg/L, the dosage is 0.5 mg/L (0.7 – 0.2 = 0.5).

Only a portion of the bulk quantity of chemical additive contributes to the optimal fluoride level. The fluoride products as delivered are not 100 percent pure, and the fluoride ions may be bound to other elements. Table 4-1 lists examples of purity and available fluoride ion (AFI) for the fluoride products discussed in chapter 3. The percent available fluoride is the purity multiplied by the AFI.

The feed rate can then be calculated using the dosage, process water flow, purity, and available fluoride ion for the fluoride product that will be used. With these data, the chemical feed rate (in pounds per day) can be calculated for the various methods of feeding fluorides by the following relationship:

$$\text{Chemical feed rate (lb/day)} = \frac{D \times Q(\text{mgd}) \times (8.34)}{\text{AFI} \times P}$$

Where:

D = Dosage in mg/L

Q = Facility flow rate in mgd

AFI = Available fluoride ion, decimal fraction

P = Fluoride product purity, decimal fraction

The 8.34 conversion factor allows milligrams per liter and million gallons per day to equal pounds.

Example Feed Calculations

Sodium Fluorosilicate. The recommended fluoride level is 0.7 mg/L with a natural fluoride level of 0.2 mg/L. The chemical supplier has indicated that the commercial purity of the sodium fluorosilicate supplied is 98 percent and that a cubic foot of the supplied material weighs 65 pounds. If the plant rate is 2,000 gpm, determine the chemical feed rate in pounds per minute, cubic feet per hour, and grams per minute.

Given:

Recommended fluoride level = 0.7 mg/L

Natural fluoride level = 0.2 mg/L

Chemical purity = 98 percent

Pumping rate = 2,000 gpm or 2.8 mgd

Chemical weight = 65 lb/cubic foot

Solution:

a. Determine fluoride dosage:

$$\begin{aligned} \text{dosage (mg/L)} &= \text{optimal fluoride level (mg/L)} - \text{natural fluoride level (mg/L)} \\ &= 0.7 - 0.2 \\ &= 0.5 \text{ mg/L} \end{aligned}$$

b. Determine available fluoride:

From Table 4-1, sodium fluorosilicate has 60.7 percent available fluoride ion.

c. Calculate feed rate:

$$\text{Chemical feed rate (lb/day)} = \frac{D \times Q(\text{mgd}) \times (8.34)}{AFI \times P}$$

$$\begin{aligned} \text{Chemical feed rate (lb/day)} &= \frac{0.5 \text{ mg/L} \times 2.8 \text{ mgd} \times 8.34 \text{ lb/gal}}{0.607 \times 0.98} \\ &= 19.6 \text{ lb/day} \end{aligned}$$

d. Convert chemical feed rate to cubic feet per hour = 19.6 lb/day

$$19.6 \text{ lb/day } (24 \text{ hr/day} \times 65 \text{ lb/ft}^3) = 0.013 \text{ ft}^3\text{/hr}$$

Convert to grams per minute:

$$\begin{aligned} \text{feed rate (g/min)} &= \frac{19.6 \text{ lb/day} \times 454 \text{ g/lb}}{24 \text{ hr/day} \times 60 \text{ min/hr}} \\ &= 6.18 \text{ g/min} \end{aligned}$$

Fluorosilicic Acid. The recommended fluoride level is 0.7 mg/L with a natural fluoride level of 0.1 mg/L. The chemical supplier has indicated that the commercial purity of the fluorosilicic acid to be fed is 23 percent (10 lb/gal). If the plant rate is 4,000 gpm (5.76 mgd), determine the chemical feed rate in pounds per minute, gallons per minute, and milliliters per minute.

Given:

Recommended fluoride level = 0.7 mg/L
Natural fluoride level = 0.1 mg/L
Chemical purity = 23 percent
Pumping rate = 4,000 gpm or 5.76 mgd
Fluorosilicic acid solution density = 10 lb/gal

Solution:

a. Determine fluoride dosage:

$$\begin{aligned} \text{dosage (mg/L)} &= \text{optimal fluoride level (mg/L)} - \text{natural fluoride level (mg/L)} \\ &= 0.7 - 0.1 \\ &= 0.6 \text{ mg/L} \end{aligned}$$

b. Determine available fluoride:
From Table 4-1, fluorosilicic acid has 79.2 percent available fluoride ion.

c. Calculate feed rate:

$$\text{Chemical feed rate (lb/day)} = \frac{\text{D} \times \text{Q(mgd)} \times (8.34)}{\text{AFI} \times \text{P}}$$

$$\begin{aligned} \text{Chemical feed rate (lb/day)} &= \frac{0.6 \text{ mg/L} \times 5.76 \text{ mgd} \times 8.34 \text{ lb/gal}}{0.792 \times 0.23} \\ &= 158.2 \text{ lb/day} \end{aligned}$$

d. Convert chemical feed rate to gallons per hour

$$\begin{aligned} &= \frac{158.2 \text{ lb/day}}{24\text{hr/day} \times 10 \text{ lb/gal}} \\ &= 0.66 \text{ gal/hr} \end{aligned}$$

e. Convert feed rate to mL/min:

$$\begin{aligned} \text{Feed rate (g/min)} &= \frac{0.66 \text{ gal/hr} \times 3{,}785 \text{ mL/gal}}{60 \text{ mph}} \\ &= 41.6 \text{ mL/min} \end{aligned}$$

Sodium Fluoride. The recommended fluoride level is 0.7 mg/L with a natural fluoride level of 0.3 mg/L. The chemical supplier has indicated that the commercial purity of the sodium fluoride to be fed is 95 percent. If the plant flow is 520 gpm (0.75 mgd), determine the chemical feed rate in pounds per day, gallons per day, and milliliters per minute.

Given:

Recommended fluoride level =	0.7 mg/L
Natural fluoride level =	0.3 mg/L
Chemical purity =	95 percent
Pumping rate =	520 gpm or 0.75 mgd
Sodium fluoride available fluoride ion =	0.452
Saturated sodium fluoride solution concentration =	40,000 mg/L

Solution:

a. Determine fluoride dosage:

$$\begin{aligned}\text{Dosage (mg/L)} &= \text{optimal fluoride level (mg/L)} - \text{natural fluoride level (mg/L)}\\ &= 0.7 - 0.3\\ &= 0.4\ \text{mg/L}\end{aligned}$$

b. Determine available fluoride:

From Table 4-1, sodium fluoride has 45.2 percent available fluoride ion.

c. Calculate feed rate:

$$\text{Chemical feed rate (lb/day)} = \frac{D \times Q(\text{mgd}) \times (8.34)}{AFI \times P}$$

$$\text{Chemical feed rate (lb/day)} = \frac{0.4\ \text{mg/L} \times 0.75\ \text{mgd} \times 8.34\ \text{lb/gal}}{0.452 \times 0.95}$$

$$= 5.8\ \text{pounds sodium fluoride per day}$$

d. Convert chemical feed rate to milliliters of solution per hour. One pound is equal to 454 grams and 1,000 milligrams equals 1 gram.

$$\begin{aligned}&= 5.8\ \text{lb sodium fluoride/day} \times 454\ \text{g/lb} \times (1{,}000\ \text{mg/1 g})\\ &= 2{,}633{,}200\ \text{mg sodium fluoride per day}\end{aligned}$$

Because a saturated sodium fluoride solution has 40,000 mg/L of sodium fluoride, dividing the required feed rate by the solution concentration results in volume to feed. A saturated solution is a solution with the maxium concentration of fluoride sodium present.

$$= \frac{2{,}633{,}200\ \text{mg sodium fluoride per day}}{40{,}000\ \text{mg sodium fluoride per liter}}$$

$$= 65.8\ \text{liters solution per day}$$

Because there are 1,000 milliliters in a liter, and 1,440 minutes in a day, the millimeters per minute that will provide the desired dosage can be calculated.

$$= \frac{65.8 \text{ liters solution per day} \times 1{,}000 \text{ mL/L}}{1{,}440 \text{ min/day}}$$

$$= 46 \text{ mL/min}$$

Sodium fluoride using a saturator:

Presuming that this sodium fluoride problem uses a saturator, there is a simplified method for the operator to calculate the pumping rate. Because saturators have a saturated solution of known concentration, 40,000 mg/L with the fluoride concentration being 18,000 mg/L (fluoride is 44 percent of the weight of sodium fluoride), the dosage can be based on the ratio of the desired dosage to the saturated solution.

$$\text{Ratio} = \frac{\text{Plant capacity (gal/day)} \times \text{dosage (mg/L)}}{\text{Solution feed rate (gal/day)} \times \text{saturated fluoride solution (mg/L)}}$$

$$\text{solution feed rate (gal/day)} = \frac{\text{Plant capacity (gal/day)} \times \text{dosage (mg/L)}}{\text{Saturated fluoride solution (mg/L)}}$$

And including the conversion of 3,785 milliliters per gallon and 1,400 minutes per day

$$\text{solution feed rate (mL/min)} = \frac{750{,}000 \text{ gal/day} \times 0.4 \text{ mg/L} \times 3{,}785 \text{ mL/gal}}{18{,}000 \text{ mg/L} \times 1{,}440 \text{ min/day}}$$

$$= 44 \text{ mL/min}$$

This page intentionally blank.

Chapter 5

Design, Equipment, and Installation

The previous chapter provided an understanding of the type of fluoride systems and the general basis of a system. This chapter will provide a more detailed perspective of the elements that comprise a fluoridation system, including the chemical storage, equipment and material handling, feeder location, and appurtenances. Finally, this chapter will include the considerations for implementing fluoridation systems. It should be noted that this same information applies to inspection and validation of older facilities that need review.

SATURATORS

Sodium fluoride has the characteristic of being easily dissolvable to a saturated solution within five minutes. The saturation concentration is nearly uniform within the temperature range that will be experienced during operation at a water treatment facility. With a nearly uniform saturation concentration across those temperatures at which most water is treated, sodium fluoride and the associated sodium fluoride saturator require minimal operator input to yield consistent concentration of fluoride from the saturator. The process involves dissolving the salt into a saturated solution and then feeding that solution in a measured rate to the facility product flow.

The cost for the equipment and installation is often the least expensive configuration, it is easy to maintain, and operation is simpler than other systems. The downsides include the cost of the sodium fluoride because it is more expensive than the other fluoride products due to the greater processing required to derive the product, and its limitations when using a high-hardness feedwater. The use of sodium fluoride in locations with high natural water hardness is a limitation as the high calcium in the water can suppress the saturated solution concentration and result in increased nuisance scaling.

Sodium fluoride has a consistent solubility of approximately 4 percent (40,000 mg/L as sodium fluoride, providing 18,000 mg/L as fluoride ion). This allows a saturator system to eliminate the need for weighing sodium fluoride, measuring the volume of solution water, and stirring to dissolve the chemical. An operator simply needs to replenish the

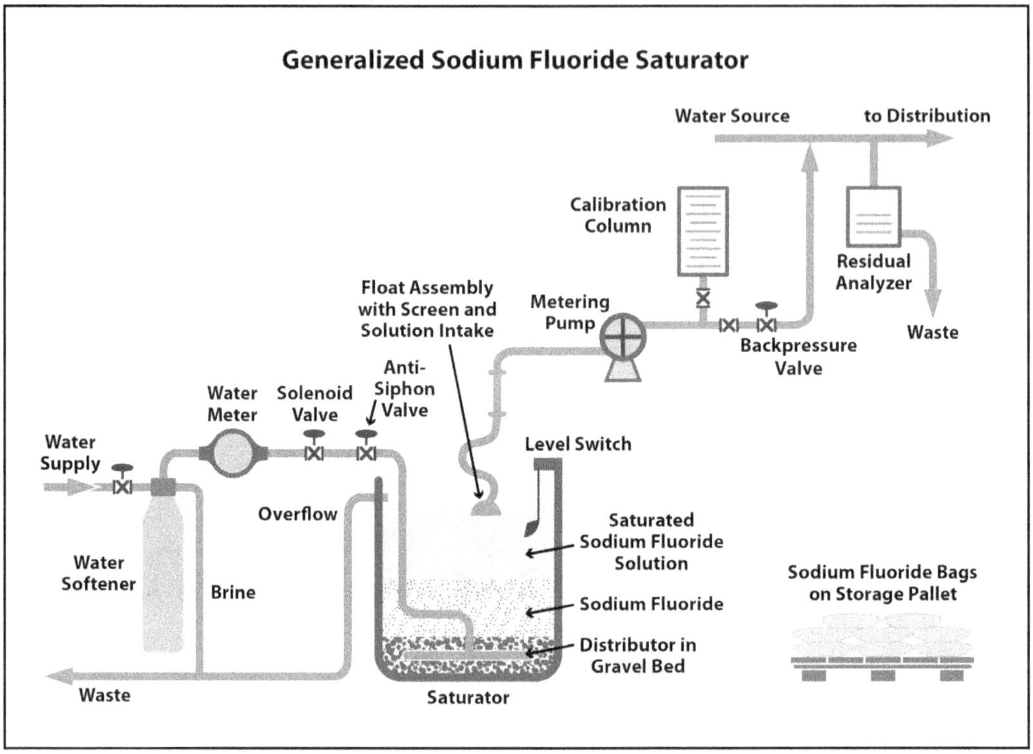

Figure 5-1 Upflow saturator

sodium fluoride salt in the saturation bed, and the makeup water will dissolve the product to a saturated solution for feeding. The process control required is simply setting the correct pumping rate for feeding the prepared solution to the plant flow injection point.

Saturators were the first application for feeding fluoride in water facilities, and were also originally used to prepare several other chemical feed systems in use at water facilities. Historically, several types of saturators were used, but at the current time all saturators are of the upflow saturator design. Saturator systems include a makeup-water connection, the saturator tank, and the solution feed system, including a pump and the necessary controls to ensure the correct and safe delivery of the solution to the plant flow injection. Saturators are typically the size of a 55-gallon drum, although a 110-gallon size is available. A bed of undissolved sodium fluoride is maintained at the bottom of the unit, through which water is forced upward under pressure using a spider-type water distributor located at the bottom of the tank (Figure 5-1). The distributor arms are typically fabricated using PVC pipe with numerous very small slits. Water, forced under pressure through these slits, flows upward through the sodium fluoride bed at a controlled rate to ensure sufficient time to obtain a saturated 4 percent solution. The water distributor slits are generally self-cleaning, with little accumulation of insolubles and precipitates. Periodic cleaning, however, is still required. The frequency of use, the rate of debris accumulation, and the hardness of the incoming water dictate the frequency of cleaning.

Saturators typically operate with sodium fluoride beds of between 12 to 18 inches. Deeper beds tend to experience uneven flow within the bed so that correct saturation may not be achieved (because some flow will bypass the bed through forced flow channels caused by the higher pressure required to enter the bed). The saturated solution is a clear layer in the upper portion of the saturator where the feeder pump intake line floats to prevent the withdrawal of undissolved sodium fluoride from the lower levels.

The feed makeup water normally includes a backflow preventer to ensure that the contents of the saturator do not contaminate the supply, a water meter to measure water consumption, and a solenoid valve to control the rate of flow to make up the saturator solution. A float in the saturator reservoir controls the solenoid valve. Most saturators operate well with ambient water as the concentration of fluoride, calcium, and magnesium ions in normal drinking water is too dilute to precipitate. However, the concentrated fluoride ions in a saturated solution can occasionally combine with elevated calcium and magnesium ions typical in hard waters and have the potential to form a precipitate that can clog the feeder, injection port, feeder suction line, or saturator bed. A commercial zeolite water softener is important if the feedwater has a hardness of greater than 75 mg/L due to the potential for the calcium to suppress the full dissolution of the sodium fluoride. Alternately, a reverse osmosis unit could be used to obtain a water purity that is suitable for use in dissolving sodium fluoride.

The actual process control is in the solution feed line. Normal components of a solution feed line would include the metering pump and a floating intake assembly, a backflow valve to ensure positive pressure delivery, and the injector.

Sodium fluoride solutions are near neutral pH, generally a pH of 7.6, so the saturator tank is not normally considered an aggressive chemical environment. The saturator tank is typically constructed of polyethylene so that it does not rust and is easy to handle. Most common elastomers are appropriate for seals and gaskets. Wetted piping can be PVC or polyethylene.

DRY FEEDERS

A dry feeder meters a chemical product at a predetermined rate. Volumetric dry feeders deliver a precise volume and are simpler to operate, are less expensive, deliver smaller quantities, and are slightly less accurate. Gravimetric dry feeders include a scale to deliver a precise weight of product and are capable of delivering large quantities in a given time period, are more accurate, are more expensive, and are readily adapted to recording and automatic control. The choice between a volumetric or gravimetric feeder is governed largely by size and economics. The selection of one of the types of dry product feeders involves several factors, such as size, cost, accuracy, and personal preference.

Dry feeders are most commonly used for sodium fluorosilicate; they can be used for sodium fluoride, although this is an uncommon application as sodium fluoride is more expensive than sodium fluorosilicate. Unlike sodium fluoride, which readily forms a consistent saturation in a rapid time, a nonsaturated sodium fluorosilicate solution has a temperature-dependent concentration. The nonsaturated sodium fluorosilicate solution has difficulty achieving full saturation, often requiring a long time period of mixing to achieve saturation. Because a saturated solution is unrealistic, the strategy with dry feeders is to use the feeder as the process control point and feed dilute unsaturated solutions to the plant product water injection. Often, dry feeders and unsaturated solutions of sodium fluorosilicates are fed to the injector using an eductor as a simple and reliable feed system, although some systems use a feed pump in place of an eductor (Figure 5-2).

Volumetric dry feeders are available in several types that differ in terms of the method of measuring and delivering the dry chemical. The two general types of gravimetric dry feeders are those based on loss in weight of the feeder hopper and those based on the weight of material on a section of moving belt. Many gravimetric dry feeders also incorporate some of the features of volumetric feeders, such as a rotary feed mechanism between the hopper and the weighing section or a mechanical vibrator for moving chemicals out of the hopper. Because it is the weight of material per unit of time that is measured and regulated, such variables as material density or consistency have no effect on feed rate. This accounts for the accuracy of these feeders. The stream or ribbon of dry material discharged from a dry feeder falls into a dissolving chamber where the dry chemical is mixed with water. The solution either drains into an open flume or clearwell or is pumped

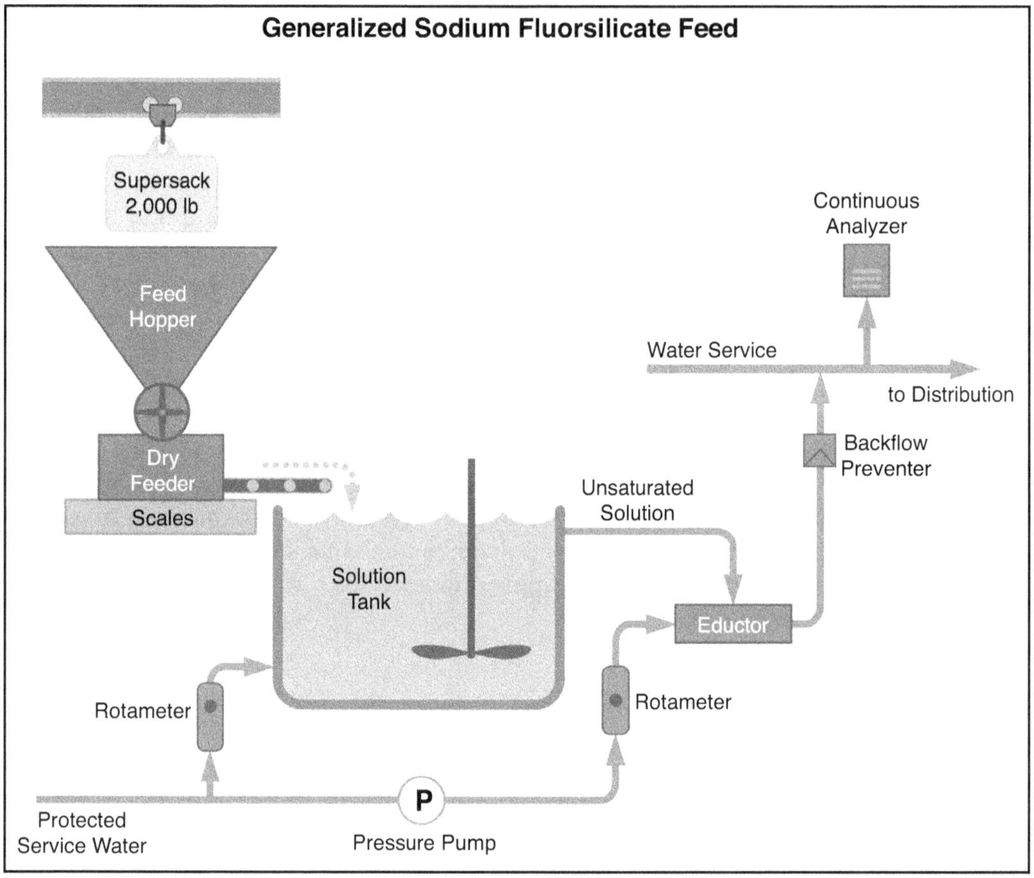

Figure 5-2 Sodium fluorosilicate feed system

or injected into a pressure main. Volumetric feeders are normally installed on weight-of-loss scales to provide recorded consumption of fluoride product used.

Feeder Size

Chemical feed equipment requires precise and reliable delivery of the product. This is best accomplished when the delivery rate is near midrange. Once the required feed rate is established, the manufacturers' bulletins and direct technical assistance should be used in the selection of appropriate equipment for a particular application. Some dry feeders have been installed for use on water supplies that are not correctly matched for the feed range, so that the feeder is operating at minimum reliable rate. In this case, the feeders are often operated below the minimum rate through the use of a cycle timer that can lead to inaccurate feed rates. Accurate and uniform fluoride feed rates cannot be expected under these conditions. It may be possible to consider replacing the delivery screw or implement with a different model so that the feed rate can be extended to the lower range.

Material Quality

Dry feeders are designed to handle crystalline materials, but not all such materials are handled with equal ease. Material that is too fine (like a powder) can flow like a liquid through the measuring mechanism and cause flooding. Some materials can form an arch (or bridge) in the hopper and, when the arch collapses, emit a cloud of dust that can flood the feeder. Other powders, either hygroscopic (water attracters) by nature or produced with a significant moisture content, tend to form lumps that can reduce the feed rate or prevent the powder from being fed at all. For most consistent feeding, a narrow crystalline size gradation and a low moisture content are best. It is important to verify that the chemical product provided by the supplier meets the correct AWWA specification size gradation for crystalline character to avoid these material handling problems.

SOLUTION DISSOLVING TANKS

Fluoride solutions must be homogeneous. Slurries should not be tolerated in the feeding of fluorides because undissolved fluoride compounds can eventually dissolve and cause a higher than optimal concentration. Undissolved material can also clog feeders and other devices that have small openings. Considerable waste can result because of the accumulation of undissolved materials.

The dry material discharged from a volumetric or gravimetric feeder is continuously dissolved in a chamber to the side of the feeder. From there, the clear solution drains or is pumped into the water to be treated. This chamber, referred to as the solution pot, dissolver tank, solution tank, or dissolving chamber, may be a part of the feeder or a separate device. Fluorides should not be fed directly into flumes or basins without the use of a dissolving tank. Use of a dissolving tank minimizes the buildup of undissolved dry chemical and makes it possible to maintain accurate feed rates.

The failure to produce a clear, homogeneous solution discharge from the dissolving tank of a dry feeder indicates one of the following problems:

- The dissolving chamber is too small, resulting in insufficient detention time.
- Insufficient solution water is provided. Because the objective is to provide an unsaturated solution of sodium fluorosilicate, sufficient dilution water should be added to obtain a solution that has no more than one-quarter of the solubility of roughly 0.762 grams per 100 mL.
- Agitation is insufficient.
- The dry chemical is short-circuiting the tank's designed route and is not being adequately mixed with the water.

Experience shows that sodium fluorosilicate should remain in the dissolving tank a minimum of five minutes. This provides a concentration of one-fourth the maximum solubility if the water temperature is greater than 60°F (16°C) and the fluoride product is a crystal gradation. If the water temperature is less than 60°F (16°C) or there is a high hardness of the dilution water, the dissolving time should be doubled; if both, the time should be tripled to 15 minutes.

Short-circuiting occurs when the travel time through a tank is less than the designed time. This reduced detention time is essentially a function of the dissolving tank design and is more likely to occur in smaller tanks. If short-circuiting does occur, baffles should be added to the tank so that the path of the chemical to the outlet of the chamber is sufficiently winding to provide the necessary detention time for the solution.

A water inlet to the dissolving tank, which is below the outlet, may create a cross-connection that requires adequate safety measures. If the dissolving tank is not already equipped with a correctly placed vacuum breaker and backflow preventer, one should be installed in the water, which delivers a specific volume of liquid. Chemical feed pumps are chosen for their ability to deliver an accurate and reliable quantity of a solution. If the water line has a solenoid or manually operated valve, a vacuum breaker must be installed between the valve and the tank for adequate cross-connection protection.

FLUOROSILICIC ACID FEED SYSTEMS

Fluorosilicic acid is the most commonly used system for adding fluoride to drinking water. Data reported by states to the CDC (2015) indicated that in 2010 over 80% of all fluoride added to drinking water was in the form of fluorosilicic acid. The largest factor in the growth of fluorosilicic acid use is the industry trend to shift to liquid feed systems. This liquid feed system avoids the need to handle dry products and the associated handling effort of mixing dry products to create solutions for the feed. A secondary factor is that hard waters can be a hindrance in mixing dry products (although the acid consumes alkalinity so there is no limitation of its use in hard waters).

Fluorosilicic acid feed systems in the most elementary form require a tank and a pump, but the potential for an overfeed mandates safety backup to prevent excessive feed of the acid. The minimum allowable system would include a bulk storage tank; a manually controlled transfer pump used to transfer the solution to a day tank installed on a platform scale; a precision solution feed positive displacement pump; a flow measurement device to verify solution feed rate using either a calibration column or flowmeter; and backflow, siphon valve, and backpressure valve. Many states have specific criteria for feeding fluorosilicic acid, and the state drinking water administrator should be consulted on system requirements for installations.

METERING PUMPS

Fluoride must be added in precise proportion to the quantity of water being treated, and precision metering pumps are necessary. A metering pump must deliver fluoride with accuracy and uniformity because the optimal fluoride level is prescribed between very narrow limits. Fluoride solutions are a particular challenge due to the very small quantities that are desired and the importance of a precise delivery so that excessive solution is not delivered.

Metering pumps essentially create positive displacement; they operate on the principle that faster speed correlates to higher delivery. Although any metering pump could be considered, pragmatically, only two types are typically used for fluoridation: diaphragm pumps and peristaltic pumps.

- Diaphragm pumps operate by a pulsing diaphragm, and the rate of delivery can be varied by both the stroke (distance that the diaphragm pulses) and the number of strokes per minute. This results in a very large effective range in capacity. Most diaphragm pumps used for fluoridation have a flexible diaphragm driven by a mechanical linkage. An electronic pump has a solenoid that is periodically energized to move the flexible diaphragm. It has solid-state electronics, circuit breakers, and an anti-siphon valve. It can have eight manual or automatic controls and variable stroke and speed. A diaphragm pump can achieve higher discharge pressures when that is an important criterion.
- Peristaltic pumps operate by a rotating element that squeezes a hose or tube as it rotates, forcing a defined volume to move in the hose/tube. They have several advantages over diaphragm pumps because the check valves, diaphragm, and backsiphonage devices are all eliminated and maintenance is simplified to changing the hose/tube.

Solution feed pumps will have their most accurate feed rates in the middle of their operating range. This is particularly true for the lower end of the delivery range where a small pulse of chemical is followed by an extended period of no delivery. The mixing of an irregular delivery is a challenge to a consistent mixing in the plant product water. It is typically better to provide two smaller pumps than a single larger pump so that dosing at the lower levels can be more consistent and uniform.

Transfer pumps are useful when transferring hydrofluorosilicic acid from storage to the day tank. A facility may want a transfer pump if more than one chemical storage tank is provided. Centrifugal pumps are most commonly used because they can run continuously and can be throttled without damage. It is essential that a transfer pump is not automatically controlled; it must be manually activated and operated under supervision. This will prevent the potential for overfeeding fluorosilicic acid through an unattended operation.

Auxiliary Equipment and Appurtenances. Auxiliary equipment is necessary for the proper operation of a fluoride chemical feed system. The type and complexity of the equipment required depends on the size and complexity of the fluoridation system. At a minimum, the required auxiliary appurtenances are limited to those needed for safe and reliable operation, and system complexity can increase as automation and control increases in complexity.

FLOWMETERS

Flow measurement is required for the accurate control of fluoride solutions. For solution mixing with a sodium fluorosilicate, the flow measurement can be an indirect measurement of the makeup water to verify sufficient dilution water from an unsaturated solution. Sodium fluoride saturators and fluorosilicic acid are a direct measurement of the solution feed.

Flow meters are matched to pipe size, not flow rate. All too often, the meter in use is oversized for the rate of flow going through it. Because most meters are least accurate at the low end of their measurement range, the result of this oversizing is that water flow is not accurately measured. A meter no larger than that necessary to handle the maximum flow rates expected should be used, even if the pipe and meter sizes do not match. In some cases, this may involve the use of pipe reducers to adapt the water main to the suitable meter size.

In a facility where the water output is variable, such as when more than one pump is used, a flowmeter serves two purposes. First, the flowmeter indicates the flow rates for which the fluoride feeder or feeders must operate. Second, if so designed, flowmeters provide an electrical, pneumatic, or hydraulic signal that adjusts the device output to correspond to changes in water flow rate (this is known as flow pacing).

Some flow measurement may be manual, for example when using a rotameter. A rotameter will not provide a remote indication on a display, and cannot control the pump delivery. However, a rotameter can be used to measure actual flow on an instantaneous basis. Since fluorosilicates are acidic solutions, it is necessary to ensure that a flowmeter is constructed of materials suitable for the acid service condition.

SCALES

Scales are necessary for weighing the quantity of fluoride product used in preparing the solution, the quantity of solution fed, or the quantity of fluoride compound or fluorosilicic acid delivered by the appropriate feeder.

The most generally applicable type is the platform scale, on which a solution tank, a carboy of acid, or an entire volumetric dry feeder can be placed.

No particular problems should be encountered when equipment is being mounted on platform scales unless the equipment is already connected to a water line or the discharge line from the dissolving chamber of a volumetric dry feeder is fixed in place. In these situations, the dissolving chamber (solution pot) will also need to be mounted on the scale platform, and all connections to it must be flexible enough to permit the scale to operate.

STORAGE OF BULK FLUORIDE PRODUCTS

Fluoride products are generally delivered in a bulk form, either as pallets with fifty 40-pound bags of sodium fluorosilicate and sodium fluoride; 2,000-pound supersacks of sodium fluorosilicate; or fluorosilicic acid deliveries in 20,000-gallon rail tank cars, 5,000-gallon trucks, 500-gallon mini-bulk, 300-gallon tote tanks, 50-gallon drums, and 20- or 13-gallon carboy tanks.

The dry crystalline products should be stored inside a building protected from the weather in a segregated room, with no mixing of other chemical products, and on pallets to keep the bags off the floor where they have greater potential of absorbing moisture. The storage room should be properly ventilated for dust control in accordance with applicable safety codes. The handling of powdered, dry chemicals can raise dust. When the quantities of fluoride compounds are small, good housekeeping and handling practices will minimize the dust problem. When larger quantities (more than one bag at a time) are handled, dust prevention and collection facilities should be provided. A dust canopy enclosing the hopper-filling area and equipped with negative ventilation to pull in any dusting will prevent the spread of dust throughout the loading area. If the plant location is congested, dust filters can be incorporated into the exhaust system.

Fluorosilicic acid can be stored outside as long as it is protected from freezing. If the tank is to be housed inside a building, then the tanks should be correctly vented to the outside as there will be a slight volatilization of hydrogen fluoride during storage, resulting in the potential for severe corrosion to interior structures. Fluorosilicic acid is nominally less than 0.25 percent hydrogen fluoride, but since hydrogen fluoride is volatile at room temperature, any gas that reaches the surface will volatilize. The bulk storage tank should preferably be polyethylene construction, which is the best material for containing fluorosilicic acid. An alternate can be a fiberglass storage tank constructed with a double-synthetic veil on the interior with a methyl ethyl ketone peroxide (MEKP)–mediated postproduction curing to minimize the susceptibility of corrosion of the glass strand reinforcement. The feed line to the tank should be equipped with a filter-strainer to strain coarse garbage and debris that is common in fluorosilicic acid deliveries so that it does not enter the bulk storage tank.

The storage area for fluorosilicic acid as well as the handling areas should have firm containment of 110 percent of the storage volume. All concrete surfaces should be coated for corrosion control using an epoxy undercoat and a polyurethane topcoat.

DAY TANKS

When fluorosilicic acid is used, a day tank is essential for safety in avoiding overfeeds. The term *day tank* is sometimes used in chemical handling schemes to refer to a convenient supply close to a point of use. For fluoridation using fluorosilicic acid, however, the day tank is strictly for a safe and reliable prevention of overfeed of fluoride product. Fluorosilicic acid is normally stored in larger tanks than those used for a daily feed, and that large quantity of chemical presents a risk of overfeed. By using a day tank between the larger storage tank and the solution feed pump, that massive risk of overfeed is eliminated. Historical experience with fluorosilicic acid installations has proven the necessity of such a prevention strategy.

The day tank (or carboy) containing the acid (if stored indoors) must be vented to the outside atmosphere, and all connections must be sealed to the outside of the building to vent the volatile hydrogen fluoride. The day tank should be positively filled by pumping on manual control with supervision and should not be automatically filled by level control or its purpose will be defeated. The day tank should be at an elevation higher than the bulk tank so that if it overflows during refilling, it will overflow back to the bulk storage tank.

A physical day tank is the most reliable method to ensure overfeed protection, but there is a trend to rely on electronic devices to replace the need for a physical day tank. An "electronic day tank" system requires two independent devices so that if one device fails, the remaining device can trigger a response to an overfeed. Although various devices could be configured, some choices include residual analyzers to measure the fluoride level, scales to measure the loss of weight coupled to a programmable logic controller to activate if the weight loss exceeds a pre-established threshold of fluoride added, a flowmeter measuring the volume of solution added, or another suitable device. If the device triggers a threshold criterion, the fluoride feed would be terminated and an alarm indicated.

BAG LOADERS

When the hopper of a dry feeder is located directly above the feeder—that is, when the operator must lift the chemical to a considerable height in order to fill the hopper—a bag loader is more a necessity than a convenience. A bag loader is essentially a hopper extension large enough to hold a single bag of chemical. The front of the loader is hinged so that it will swing downward to a more accessible height. The operator places a bag in the hinged section, opens the bag, and then swings the section back into position. This device not only lessens the labor required to empty a bag, but it also contains dust created from emptying bags. Including negative ventilation within the enclosure will further minimize dust release.

BACKFLOW PREVENTION DEVICES

Any time potable water is connected to a chemical solution tank, the possibility of a cross-connection exists. This can occur in the supply line to a saturator or dissolving tank or in the discharge from either of these or any solution feeder.

The simplest method for preventing potential siphonage is to provide an air gap in the water feed system. When pressure is high enough to make an air gap impractical, a reduced pressure principle backflow preventer or a double-check valve backflow preventer should be used. For the highest level of safety, the device should be installed as close to the chemical solution tank as possible. The use of mechanical vacuum breakers is not recommended because of the potential siphonage hazards that can exist during an instantaneous pressure loss. Refer to AWWA Manual M14, *Recommended Practice for Backflow Prevention and Cross-Connection Control*, for additional guidance.

PIPING AND VALVES

Appropriate materials for piping and valves include polyethylene (PE), polyvinylchloride (PVC), chlorinated PVC (CPVC), and polyvinylidene fluoride (PVDF). Polyethylene pipe has a large advantage due to its property of having heat-fused joints so that potential leaks are minimized. PVC is the most commonly used product due to its ease in solvent welded joints, but it is prone to leaks at poorly assembled and incompetent joints or damaged connections. PVC pipe should have rigorous pressure testing prior to use and requires periodic inspection for joint leaks as the joints all represent potential leak locations. Polyethylene and polypropylene tubing can be used for small metering pump suction and discharge lines. If pipe is to extend outside a building to the point of injection, it is important to use a pipe casing to contain potential leaks and ground surface markings to safely identify the pipe location. The flow direction and the chemical carried should be clearly marked on all chemical service piping. In addition, chemical service piping should be color coded in accordance with the requirements of the state or local drinking water regulatory entity. Great Lakes – Upper Mississippi River Board of State and Provincial Public Health and Environmental Managers *Recommended Standards for Water Works* (10 States Standards) suggests a color coding for fluoride of light blue with red bands.

PVC is the most commonly used pipe for fluoride solutions, but it also has a higher propensity for joint leaks over polyethylene pipe with integral heat fusion welded joints. It is essential that all PVC piping be pressure tested at least once every two years for identification of weak joints or other weak locations requiring replacement. Pressure testing is accomplished by using an air compressor to apply 5 psi over the pressure rating of the piping system for a prescribed time interval. If PVC is used for the fluoride-solution piping, experience has shown that numerous unions should be incorporated into the piping so that when leaks are experienced, the pipe repair is facilitated by minimizing the length and sections that require replacement.

CORROSION CONTROL

Sodium fluoride solutions have a pH-neutral character and can be used with most common products used in water treatment facilities. Sodium fluorosilicate is a weak acid solution but fluorosilicic acid is a strong acid and chemically aggressive toward many materials. PE and PVC are both compatible with fluoride solutions, but polypropylene, rubber-lined steel, or Kynar-lined steel tanks are acceptable. Fiberglass can be used with qualification in that the tanks need periodic inspection for integrity of the gel coat due to glass reinforcement being subject to attack. Unless a water facility has the ability to provide that specialized inspection, fiberglass should be avoided. Fiberglass tank construction should use a double synthetic veil on the interior and an MEKP mediated postproduction cure. Most metals are potentially subject to chemical attack and should be avoided. Steel tanks can be used if rubber-lined, but they require inspection and repair of the rubber lining periodically as the liner will eventually exfoliate. The nickel in stainless steel is susceptible to fluoride exposure and the welds are vulnerable (because stainless welds have high nickel content). If concrete is used as a base for the tank or as a chemical containment barrier, then it should be coated to resist minor exposure using an epoxy undercoat with a urethane topcoat. Alternately, a personnel access opening rehabilitation spray-on polyurethane coating could be satisfactory.

It is important to verify with equipment manufacturers the compatibility of the materials used in their products with fluoride solutions. Most will have a history of favorable compatibility and can provide references on suitable products and materials.

CONTINUOUS ANALYZERS

Continuous analyzers measure and display the fluoride content of a solution. They use a fluoride-specific electrode and normally record a discrete measurement once every interval of time. The experience with these devices has improved and many utilities report satisfied response even when the natural waters have a low salinity and total dissolved solids (TDS) content of less than 100 microSiemens/centimeter (μS/cm). A low salinity/TDS is a potential error factor as the ion specific electrode method of measurements typically requires a total ionic strength adjusting buffer to swamp the ionic background to enhance the performance of the electrode. The Analysis Methods section of chapter 6 provides additional information on the ion specific electrode test procedure.

Although the experience is that a continuous analyzer can be a reliable monitoring device once it achieves a stable operation in an installation, it does require regular calibration and verification. The device needs to be calibrated and verified at least once every week at a minimum for instrumental drift (although daily verification is better for identifying and addressing solution plugging of orifices and improperly operating electrodes). At least once each month the electrode should be inspected and calibrated for correct operation as the electrical output can degrade over time.

INJECTION LOCATION

Many states require that chemical additives be introduced at a location of positive pressure. Having a location of positive pressure for addition can also be coupled with a flow switch that registers when there is a positive flow in the line. This configuration provides additional safety in that the chemical feed pump might be activated by parallel operation to a plant or well flow pump, but if the flow switch does not detect actual flow in the line, it can prevent the addition of fluoride into a line that is not actually flowing. An injector such as a Monel alloy can provide sufficient chemical compatibility with fluoride solutions. The injector should be introduced in the lower half of the pipe and directed upward into the pipe. Because the fluoride solution is heavier than water, injecting it upward enhances the mixing with the flow; a downward injection would direct it to the pipe invert where it may remain in the plume and not readily mix with the fluid above.

CALIBRATION CYLINDERS

A calibration cylinder is a vertical plastic column with volumetric markings. Each pump should be configured with a calibration cylinder to allow measurement of the pump delivery. If fluorosilicic acid is the solution being handled, then the top of the calibration column should be connected to a vent to the outside of the building to ensure that hydrogen fluoride fumes are not released through the column.

WASTEWATER CONNECTIONS

There are two instances where a connection to a sewer might be necessary. One is if a continuous analyzer is installed with the need for waste flow, buffering, and calibration chemicals that require disposal. The second is when a zeolite water softener or reverse osmosis unit is used for reducing hardness in saturator makeup water feed. If a sanitary sewer is available at these locations, then the disposal is simplified. If no public sewerage is available, such as at a remote well site, then a septic tank would be necessary. The flows from these sources are small in volume and the chemical character is compatible with septic tanks, but the issue will be accommodating the volume of flow. The number of gallons over the course of a week will need to be established and that quantity of flow will need to be considered when sizing the septic tank and calculating the required size of the leaching field for hydraulic assimilation in the soil.

In some cases, it may be possible to divert the waste flow to a sludge disposal process—for example, if a water treatment facility is located in a remote location that has no public sewerage but does have a generation of waste sludge. Some states may allow the minor flows from water softening or the continuous analyzer as a compatible flow to the sludge processing disposal (such as a lagoon or drying bed) if there is no other reasonable disposal option.

FACILITY CONSTRUCTION AND STARTUP

A well-designed project is key to a successful project, but a properly constructed project ensures correct implementation. The contractor should be knowledgeable and experienced in the construction of water treatment facilities and be flexible in working with the staff of the existing facility where the fluoridation facility is being added. During construction, the design engineer normally conducts site visits to inspect the system installation for compliance with the contract documents, as well as the quality of construction. Once construction is complete, the system is inspected by the permitting agency for compliance with contract documents and other pertinent regulations.

The major items for inspection can be grouped into the following categories:

- Process and ancillary equipment: The type, size/capacity, and material of tanks, and the metering pumps, water softeners, valves, pipes, and so on, are in compliance with the contract documents. Some manufacturers' representatives can be available for installation confirmation and inspection. One example is that a peristaltic pump can have different tube/hose sizes that directly affect the delivery rate. Thus, it is essential to verify that the correct size of tube or hose is installed.
- Installation: Equipment has been installed according to the contract documents. For example, different pieces of equipment have been connected correctly to follow the process flow, adequate clearance has been provided to allow access to equipment, and equipment has been adequately anchored and supported and restrained for wind and/or seismic movement (where applicable).
- Structure: The structure housing the fluoridation equipment has been constructed according to the contract documents. For example, wherever specified by the contract documents, curbs have been constructed to provide chemical containment, sumps have been built to facilitate removal of spilled liquid chemical, and any required chemical-resistant coating has been applied.
- Safety, HVAC, fire protection equipment: Emergency showers and eyewashes have been installed at easily accessible locations; HVAC and fire protection equipment, if required, has been installed to comply with local building and fire codes and is in working order.
- Instrumentation and controls: Process control instruments, such as flowmeters, tank level sensors, and pressure gauges, have been installed as specified in the contract documents.

Appendix E provides inspection checklists for the three types of fluoride systems.

When the construction of the fluoridation system is mostly complete, testing is performed on individual components as well as the entire system to ensure that the system will operate as designed. Testing procedures and criteria for system acceptance are specified in the contract documents. Most testing is carried out by the construction contractor and at times with the aid of equipment suppliers. The design engineer may be present during some tests to provide technical assistance. Some jurisdictions may also require representatives of permitting agencies to be present at certain tests. It is important to keep accurate records of testing procedures and results.

The major system components that are typically tested can be grouped into the following categories:

- Pressure testing of pipes and tanks: All pipes and tanks are filled with the appropriate fluid and pressurized for a duration specified in the contract documents. Clean water is typically used for functionality testing.
- Containment area leak test: The containment area or structure within which the fluoridation equipment is located is leak tested.
- Chemical feed equipment: Gravimetric feeders are tested for accuracy with known weights. Metering pumps are calibrated, and a calibration curve is developed for each metering pump.
- Other equipment: Valves, HVAC equipment, and other ancillary equipment (e.g., sump pumps) are tested to ensure that they operate properly and reliably. Instrumentation and process control equipment is tested and calibrated.
- Safety systems: Emergency showers and eyewashes are tested to ensure the necessary flow rate is delivered. The fire protection system, if required, is tested for functionality and capacity.
- Electrical equipment: Electrical conductors, equipment, protective devices, grounding, and other components are tested as specified in the contract documents. Motors are jogged to verify the direction of rotation.
- Instrumentation and controls: An operational readiness test and a functional acceptance test are performed to ensure that the control of the fluoridation system by SCADA has been properly programmed.

Appendix E provides testing checklists for the three types of fluoride systems.

Prior to bringing the fluoridation system online, the water provider must secure approval by the permitting agency on the fluoridation system installation and the fluoride monitoring and reporting plan, and notify the public about the upcoming fluoridation implementation. To prepare for system startup, operators of the fluoridation facility attend training on equipment operation and maintenance conducted by the design engineer and equipment vendors. Following completion of operational readiness and functional acceptance testing, the fluoridation system is ready to be brought online.

On the day of system startup, a representative from the permitting agency is typically present to observe the startup process, and the design engineer is often present to provide assistance to the operators on tasks such as fluoride dose calculations, troubleshooting equipment, and control system malfunctions.

The following are important points to note when starting up a fluoridation system:

- Start early in the day. Startup activities usually require some trial and error, adjustments to fine-tune equipment and operations, and troubleshooting of equipment malfunctions. Commencing the startup process early in the day maximizes the time resources of the design engineer, permitting agency representative, and other operators who are at the treatment facility to assist in these efforts.
- Make sure fluoride analyzer equipment has been calibrated and sufficient amounts of any necessary reagents are on site.
- Start the fluoride feed at low target doses to allow ample room for equipment and control fine-tuning without risking overdosing.
- Have an accurate understanding of the hydraulic characteristics of the water flow between the fluoride feed point and the sample point, to determine the sampling strategy after each adjustment of the metering pump settings.

- If the fluoride analyzing method is temperature sensitive, be sure to adjust the sample temperature such that it is similar to the temperature of the calibrating standard.
- Maintain accurate and complete records of all testing procedures and results.

REFERENCES

Centers for Disease Control. 2015. *Water Fluoridation Reporting System*. Atlanta, GA: CDC.

American Water Works Association. 2015. M4, *Backflow Prevention and Cross-Connection Control*. Denver, CO: AWWA.

Chapter 6

Operations and Maintenance

For the benefits of fluoridation to be realized, the fluoride concentration must be maintained at or near the recommended level of 0.7 mg/L. Experience has shown that the basic reason for low or erratic fluoride levels is poor operation and maintenance. The operator is the most important factor in achieving good operating results. The best designed facility will not meet operational objectives if the operator does not operate the facility correctly. Conversely, a good operator can often achieve excellent operations even when hindered by an imperfect design. The key factors are understanding the process and ensuring those concepts are correctly implemented. This chapter will present operational strategies, process monitoring, operator safety, and maintenance.

OPERATIONAL STRATEGIES

Adjusting the fluoride level in a water supply to the recommended level is accomplished by adding the proper dosage of fluoride product on a continuous basis. Figure 6-1 summarizes the process monitoring and control scheme for fluoridation. The proper dosage of fluoride product to add to the water supply is calculated based on the difference between the target fluoride level and the naturally occurring fluoride. As the 13th most abundant element in the earth's crust, fluoride is present in practically all waters at some concentration, but normally at a level less than beneficial for oral health. Once the dosage is known, the feeders and pumps can be adjusted to deliver the correct dosage. Routine sampling, analysis, and adjustment of the dosage will ensure that the recommended level is consistently maintained. Experience and studies have shown that ±0.2 mg/L is an achievable operating range if the operator is aiming to meet 0.7 mg/L (Brown 2014).

Operational schemes for fluoridation equipment can be as varied as the systems themselves. They range from very small systems in which one person is responsible for all aspects of the operation to multidepartmental operations in which responsibilities are divided among different teams of people. In the latter case, coordination and communications within the organization must be given special attention to ensure a smooth and safe

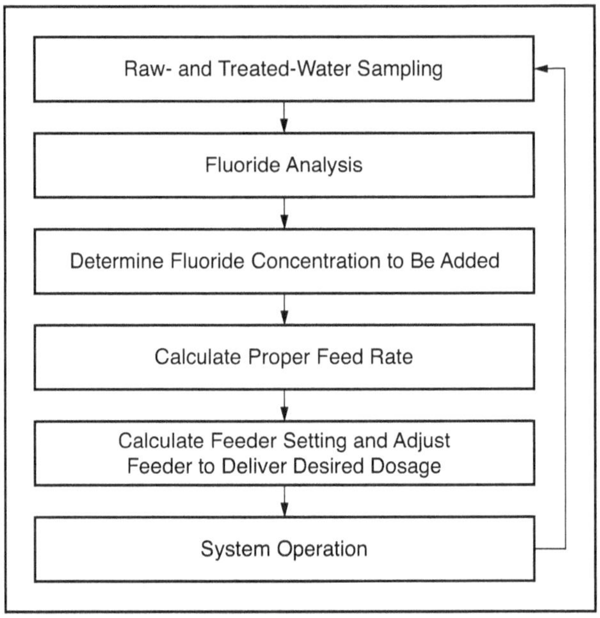

Figure 6-1 Simplified process control scheme

operation. Typically, fluoride is added daily to maintain the recommended level within the required operational tolerance, and fluoridation equipment should be inspected daily when in use. Often, the time required for this inspection will be a few minutes, but in instances that involve maintenance issues or operational problems, more time may be needed. The fluoride product should be added to a predetermined operating level for a saturator, dry feeder, or day tank as needed. Plan ahead for weekends and holidays so that a continuous feed of the chemical can be maintained. If a day tank is used, management should establish a maximum operating level to be used. Large plants might reduce the size of day tanks to fit shift changes.

A daily routine includes an inspection of the fluoride metering system, measurement and recording of the amount of chemical fed, collection of a fluoride sample at a designated sample point (usually a location that is stipulated by regulations or operating permits, and allows sufficient mixing of the fluoride solution with the water flow), and the addition of chemical to the feeder. Chemical usage and the amount of chemical added to the day tank or feeder should be maintained in a daily operator logbook along with the operator's name and date.

Each operator should be knowledgeable about the unique fill and draining characteristics of transfer lines from bulk storage tanks to day tanks at their facilities. Care must be taken to prevent siphoning from the bulk tank to the day tank after the transfer pumps are turned off. Elevation differences between the bulk tanks and day tanks can cause this problem. The transfer line should be designed to reduce this possibility and control siphoning. As previously mentioned, day tanks should be positioned above the elevation of the bulk tank so that in the event of an overfeed to the day tank, the excess returns to the bulk tank by gravity.

Dry additive feeders normally have a hopper with a sufficient supply of product for a day or more, and the correct level should be replenished daily. Excessive product in the hopper creates potential for packing or bridging in the hopper. In that case, the operator will need to check the hopper to ensure that there is free flow and the supply of additive product is able to reach the screw or impeller.

Saturators have a bed of sodium fluoride through which the water travels and dissolves the product to a saturated solution. The potential exists for the sodium fluoride bed to develop flow conduits or channels that allow water to travel through the bed without having contact with the sodium fluoride. This situation is called *piping*. To prevent this from happening and to ensure a uniform flow through the bed, the bed should be gently but firmly stirred daily with an instrument such as a stiff PVC pipe or paddle.

Standard operating procedures (SOPs) should be prepared for all operations and located on a wall near the fluoride pump or feeder. SOPs are important to remind operators of the critical activities for successful and safe operations. Operators should check with their state health department, state drinking water program, or chemical manufacturer for examples of SOPs, if available. The SOP details the daily procedures required to fluoridate at optimum levels in a safe manner. Details may include:

- Depressurization of chemical feed lines and rinsing of the metering pump before performing maintenance or repairs.
- Information on termination and startup procedures for chemical feeders, including all operations necessary to leave each piece of equipment in a safe condition.
- Specific shutdown instructions should be given for pipe segments that include chemical injection points that may be isolated from the flow for maintenance. Failing to shut off the chemical feed to a pipe that has been isolated from normal flow can allow the accumulation of acid chemical in the pipe segment. This situation can potentially damage the pipe and create a risk of a chemical release to the system when the pipe segment is placed in service.

Material safety data sheets (MSDS) are prepared by the fluoride product manufacturer. All personnel involved in fluoridation should be familiar with these sheets, and they should keep a logbook on site for every chemical product handled at the facility.

COMPLIANCE MONITORING

Regardless of whether a water system adds fluoride in treatment, fluoride is a regulated contaminant. As a result, most water systems' state or regional regulatory agency has a monitoring requirement to meet the SDWA obligation of ensuring that customers are receiving tap water with fluoride levels that do not exceed the primary federal MCL. The federal monitoring requirement for compliance with the SDWA requires surface water systems to collect a minimum of one sample annually at every entry point to the distribution system after application of treatment, or within the distribution system at a point representative of each source after treatment. Depending on state requirements, groundwater systems must collect a minimum of one sample annually or once every three years at every entry point to the distribution system representative of each well after treatment. Water system operators should consult with their state or regional regulatory agency on the requirements for annual sampling for fluoride. If naturally occurring fluoride is measured at a point to exceed the SMCL of 2 mg/L, then the monitoring frequency for fluoride is quarterly. States may waive some monitoring requirements for systems with multiple entry points by allowing for the collection of composite samples. Some states, however, may have requirements for more frequent monitoring as well. The individual water system's awareness of its regulatory monitoring requirements is important.

For a water system that adds fluoride in treatment, there may be daily or monthly monitoring requirements in addition to the annual SDWA requirements. These requirements are normally stated in operating permits, but it is important to verify reporting requirements with the state or regional regulatory agency. The sampling locations may be the same as required for SDWA compliance monitoring, or they may include different or additional locations, including locations in the distribution system. These requirements vary from state to state.

Sampling

It is important that a water system have documented procedures for collection and handling of samples. Thorough record-keeping protocols should be included as a part of these procedures. The chain of custody (COC) is used in the sample collection process to ensure samples are handled correctly and provide a hardcopy record. The COC should document the date, time, and place of each sample collected; volume collected; name of the person who collected the sample; and any ancillary notes or observations recorded by the sampler at the time of sample collection. This same information should also be recorded on the bottle itself if it will be returned to the lab and relinquished by the sample collector to another staff member for analysis. Related ancillary records might be maintained in a notebook and include such items as methods and dates for cleaning of sample bottles, inventory records of reagents, and standards used in the analysis of samples.

A *grab sample* is a sample collected at one specific location at one specific time to provide instantaneous water quality information for that specific place and time. A *composite sample* is composed of various samples either collected in one location at various times or collected from several locations and mixed together to form one combined (composite) sample. Composite samples often are used for regulatory compliance reporting, but they do not provide much useful information about specific treatment processes. Therefore, grab samples are commonly used to make process control decisions.

Standard Methods for the Examination of Water and Wastewater (Standard Methods; APHA, AWWA, WEF 2012) instructs that samples for fluoride analysis are to be collected in glass or plastic bottles. The volume collected may vary depending on the analytical method but 100 milliliters (mL) should be sufficient. No preservative is required and samples do not need to be kept cool. Although samples will typically be analyzed immediately following collection for process control purposes, samples for fluoride analysis may be analyzed up to 28 days following sample collection. This may be relevant for any samples collected and analyzed to meet SDWA monitoring requirements if the sample will be sent to an off-site laboratory. Facilities should have procedures for ensuring sample bottles are clean and do not affect the measured value of fluoride in the sample. At a minimum, sample bottles should be rinsed with deionized (DI) water.

Whether collecting the sample from a tap or a dip device, it is a good practice to rinse the sample bottle (with the water to be sampled) and discard that water before collecting the actual sample. This procedure serves to both rinse and condition the sample bottle. If the sample is to be collected from a tap that is not open and continuously running, open the tap and allow it to run for several minutes (until the water temperature stabilizes) prior to collecting the sample. This ensures that the water collected in the sample bottle is of typical quality found in the system and not water that was sitting stagnant in the plumbing.

Analysis Methods

The ability to detect and accurately measure the fluoride ion in water is essential to ensuring consumers are obtaining the benefits of fluoridation. In addition to continuous monitoring of fluoride concentration in treated water, it is also necessary to monitor its concentration in the source water prior to treatment in order to determine how much chemical must be added to reach the target levels in the finished water. When choosing an analytical method, it is important to know the target level of fluoride being added to the water to ensure the analysis method has the sensitivity required to reliably measure the

Table 6-1 Analytical methods currently approved for use for SDWA monitoring programs

Method	USEPA	ASTM	*Standard Methods* (18th through 22nd Ed., and online Standard Methods)
Ion Chromatography	300.0, 1993	D4327-97	4110B
Distillation; colorimetric analysis (SPADNS)			4500-F⁻.B, D
Ion specific electrode		D1179-93B	4500-F⁻.C

Source: 40 CFR part 141.23

concentration at approximately that level. The analyst must also be aware of components within the sample that may interfere with the measurement of fluoride.

Table 6-1 lists the analytical methods that are currently approved for use for SDWA monitoring programs in the *Federal Register* at 40 CFR part 141.23. State or regional regulatory agencies may have primacy for determining acceptable analytical methods and should be consulted to ensure acceptable analytical methods are being used. If the colorimetric or ion chromatography analytical methods are to be used, it is important to ensure that analytical interferences are not normally present in the water source.

FLUORIDE (MANUAL) ION SPECIFIC ELECTRODE (ISE) METHOD

The ion specific electrode method is based on the use of an electrode resembling the type of electrode used to measure pH. Typically 10 to 25 mL of the sample is used for analysis. The electrode measures the potential across a lanthanum fluoride crystal relative to a reference standard solution. The electrode is actually measuring fluoride activity, not concentration. The activity is less than or equal to the concentration; any difference between activity and concentration is caused by characteristics such as the solution's ionic strength, pH, and fluoride-complexing components. Use of a total ionic strength buffer (TISB) in all standards and samples provides for a uniform ionic strength between all standards and samples, adjusts pH to a level that is optimum for fluoride sensitivity, and releases free fluoride ions from complexing species. The ISE analytical method is not subject to many analytical interferences in a typical drinking water sample. The interfering substances that can introduce error to the ISE measurement are given in Table 6 2.

For most drinking water samples, these potential interferences will not be a factor in ISE measurement. The most likely constituents of a sample that may affect the electrode's measurement are elevated levels of aluminum or total dissolved solids (TDS), or a pH that is very acidic or basic, although this is not common in most drinking water samples. The use of cyclohexylenediaminetetraacetic acid (CDTA) buffer should minimize the effects on analysis caused by high levels of aluminum or extremes in pH.

The procedure involves comparing the potential of a reference standard solution with known fluoride concentration to DI water with no fluoride content. Then the measurement of fluoride in a sample of water can be compared to those reference solutions. It is important that each analyst who uses the electrode method become familiar with the instrument's operation. The meters do not typically generate one specific numerical value corresponding to the fluoride concentration in the solution. Rather, the concentration reading on the meter increases or decreases as a function of the fluoride level in the solution being analyzed, and the reading becomes more stable as the true concentration in the solution is reached. It is up to the analyst's judgment and experience to determine when the reading has stabilized enough to record the result.

Table 6-2 Interferences for ISE measurements

Substance	SPADNS	Electrode
Alkalinity	5,000 (–)	7,000 (+)
Aluminum	0.1 (–)*	2.0 (–)
Chloride	7,000 (+)	20,000 (–)
Iron	10 (–)	200 (–)
Hexametaphosphate	1.0 (+)	50,000
Phosphate	16 (+)	50,000
Sulfate	200 (+)	50,000 (–)
Chlorine	Must be completely removed with arsenite	5,000
Color and turbidity	Must be removed or compensated for	—

Notes: Concentration of substance, in milligrams per liter, required to cause error of plus or minus 0.1 mg at 1.0 mg/L fluoride. (+) causes a higher fluoride reading than actually exists; (–) causes a lower fluoride reading than actually exists.

* Figure is for immediate reading (the figure will change with time).

It is important that manufacturer-recommended guidelines for maintenance be followed when using the electrode method. In general, if an electrode responds quickly (not sluggishly) to the changing fluoride concentration in a solution and exhibits good recoveries of quality control standards, it is functioning properly. Proper maintenance (replacement of the electrode fill solution, regular cleaning of the electrode's membrane surface) is necessary to maintain a properly functioning electrode.

The electrode will lose functionality over time, so it is essential to periodically calibrate the electrode according to the manufacturer's instructions and then verify its operation. This calibration is easily performed by measuring the differential millivolt measurement over an order-of-magnitude fluoride concentration, such as 0.1 to 1.0 mg/L, or 1.0 to 10 mg/L. A correctly functioning electrode will register a 56-millivolt differential over one order-of-magnitude range. As the electrode loses effectiveness as a result of crystal mineralization or loss of integrity in the electrical lead within the electrode, the differential will decrease and the operator will need to decide when the loss of performance warrants replacement of the electrode. For routine regulatory compliance testing, some operators will accept a 90 percent effectiveness (51-millivolt differential over one-order of magnitude) as satisfactory, but they will replace the electrode when it drops below that level. A more precise tolerance (95 percent) provides more accurate measurement.

COLORIMETRIC SPADNS METHOD

The SPADNS method is a colorimetric method based on the zirconium-SPADNS dye reaction that produces a red dye lake (deep color). When dye is added to a sample containing fluoride, a portion of the fluoride in the sample reacts with the dye and forms a colorless complex; the higher the concentration of fluoride in a sample, the lower the color intensity of the dye lake. The colors produced by different concentrations of fluoride are all shades of red, making it challenging to detect those differences with the naked eye, thus requiring the use of a photometer to measure the loss of color. Approximately 50 mL of sample is used for analysis.

The procedure involves comparing the absorbance at 570 nanometers (nm) of a reference standard solution with known fluoride concentration to DI water with no fluoride content (known as a *blank*). The measurement of fluoride in a sample of water can be compared to those reference solutions. The instruments on the market provide an automatic comparison and then provide a direct measurement.

Colorimetric analysis methods tend to be subject to more interferences than other methods. Users should be aware of those interferences and their impact on results. Colorimetric analysis tends to be a less accurate method to determine fluoride content versus other methods, but many water facility operators prefer the simple procedure and equipment, and this is the most commonly applied measurement method. Proprietary methods using prepackaged vacuum ampules with the reagent draw a sufficient volume of sample to mix with the reagent, and many operators prefer the ampule procedure. Use of pipettes and cuvettes for sample handling is less expensive and results in less waste for disposal. The SPADNS reagent contains arsenic as a chlorine/oxidant scavenger, so disposal of the solutions can require special handling and processing. Some proprietary reagents have an alternate chlorine scavenger, allowing for easier disposal.

Interferences are more of a factor in the colorimetric method than in the ISE method. Table 6-3 lists common interferences in the SPADNS and ISE methods, the concentrations at which these interferences affect each method, and the type of interference (positive or negative bias) caused. The colorimetric method is more sensitive to lower concentrations of interfering substances than the ISE method. Further, the SPADNS method is affected by chlorine and turbidity, both of which must be removed prior to analysis. Any sample suspected of containing interferences should be distilled prior to analysis; Standard Methods provides the correct procedure for distillation. Also, the colorimetric method has a limited allowable range of measurement, with the sample requiring a minimum of 0.1 mg/L of fluoride and a maximum level of 1.4 mg/L. Samples with fluoride levels $\gtrsim$ 1.4 mg/L should be diluted for analysis.

ION CHROMATOGRAPHY

Ion chromatography separates a mixture of ions in a liquid sample (mobile phase) traveling through a column (stationary phase). Different ions have different migration rates through the column. The ions are separately identified by the delay in their detection as they exit the column. This method allows for measurement of several anions using the same sample at the same time. The equipment is expensive, and the technician training is significantly greater than the other tests, so it is typically a procedure for larger laboratories needing to collect numerous characterizations of multiple samples. A small volume of liquid sample (approximately 1 mL) is injected into a sample loop that will ultimately use approximately 0.1 mL of the sample for separation and analysis. The sample is passed through an anion exchange column in a stream of solvent (known as the *eluent*). The individual anions have differing affinities for the strongly basic anion exchange columns, so they travel at different speeds through the column and are separated. The anions are now in their more highly conductive acid forms, and the individual anion peaks in conductivity are identified on the resulting chromatogram based on their known retention times on the analytical column (Figure 6-2).

Fluoride's conductivity signal is the first peak on the chromatogram, and it has the shortest retention time, eluting about one minute after sample introduction to the columns. Fluoride's short retention time is a limitation that is specific to fluoride analysis using this method. The detector's response to conductivity drops below the system's baseline conductivity just prior to fluoride's elution from the column and subsequent detection. This suppression in the baseline is caused by the elution of water from the column and is referred to as the water dip. The water dip may cause difficulty in quantifying the conductivity signal caused by fluoride relative to the normal baseline. The instrument

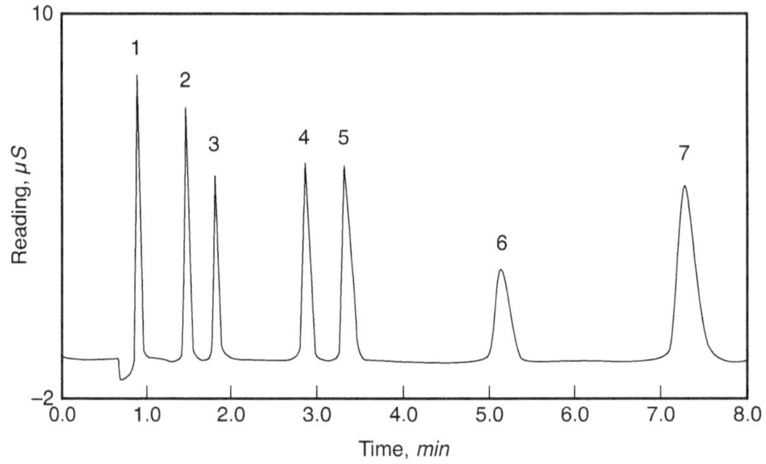

Anion	Concentration, *mg/L*
1. Fluoride	2.0
2. Chloride	3.0
3. Nitrite	5.0
4. Bromide	10.0
5. Nitrate	10.0
6. Orthophosphate	15.0
7. Sulfate	15.0

Figure 6-2 Example chromatogram

manufacturer should be able to provide recommendations about proper column/eluent combinations for minimizing the impact of the water dip.

There are other interferences, including simple organic acids that may elute at or close to fluoride's retention time and thus interfere with the fluoride peak. These small compounds (such as formate, acetate, and propionate) are not normally present in most drinking water samples, but some water sources may have seasonal occurrence so the analyst should be aware that if present in the sample, they may interfere in the determination of the fluoride concentration.

Standard Methods recommends using a dilute eluent, a gradient elution with an NaOH eluent, or alternative column types for measurement of fluoride.

RELATIVE ERROR

When testing for fluoride, the ion specific electrode (ISE) is the method with the best results and fewest interferences, while the colorimetric method has the least reliable results with the most interferences. Standard Methods presents an observed error in a referenced round-robin comparison between laboratories of ±8 percent in one standard deviation and 5 percent for ISE (Standard Methods section 4500-F^-). The CDC Laboratory Proficiency Testing Program experienced a variable error that was slightly greater than that of Standard Methods. A comparison of the ISE and colorimetric methods in the CDC Laboratory Proficiency Testing Program demonstrated the measured relative error and confirmed that the ISE is more accurate even when there are no interfering substances (Table 6-3).

Table 6-3 Measured relative error for ISE method compared to SPADNS method

	SPADNS		ISE	
True mean mg/L	One standard deviation mg/L	Data range mg/L	One standard deviation mg/L	Data range mg/L
0.5	0.040	0.46 – 0.54	0.025	0.47 – 0.52
0.7	0.056	0.64 – 0.75	0.035	0.66 – 0.73
0.9	0.072	0.82–0.97	0.045	0.85 – 0.94
1.5	0.120	1.38 – 1.62	0.075	1.42 – 1.57
	Two standard deviation mg/L	Data range mg/L	Two standard deviation mg/L	Data range mg/L
0.5	0.080	0.42 – 0.58	0.050	0.45 – 0.55
0.7	0.112	0.59 – 0.81	0.070	0.63 – 0.77
0.9	0.144	0.76 – 1.04	0.090	0.81 – 0.99
1.5	0.240	1.26 – 1.74	0.150	1.35 – 1.65

Notes:

1. These were conducted with reference laboratories participating in the CDC Proficiency Testing Program between 2006 and 2009.
2. Based on 256 referenced round-robin samples on 109 unique solutions ranging from 0.32 to 5.47 mg/L containing no interfering substances with 66 to 87 laboratories participating in the referenced rounds.
3. Insufficient number of ion chromatography data to include in comparative error analysis.
4. One standard deviation encompasses 66 percent of data; two standard deviation encompasses 95% of data.

RECORD KEEPING AND DOCUMENTATION

Comprehensive record keeping is necessary for the efficient operation and maintenance of the water supply system. The records provide an outline of required actions, their frequency, and the actions taken. This type of understanding can help new employees to effectively assume responsibility when necessary. Because the sizes and designs of water supply systems vary, record-keeping methods must be tailored to each system. Permanent records describing the operating practices in a system promote communication between staff and strengthen an operation. A well-kept system of records demonstrates how the operator maintains system reliability and provides a history of the process.

Most states require that certain records be kept on file for a certain period of time. Generally, this includes the daily determination of fluoride concentration, daily weights (or volumes) of chemicals fed, and the volume of water treated. Some states also require the daily calculated chemical dosage. Additional information, such as when the equipment was checked, repaired, and lubricated, and when chemicals were added, should also be maintained.

Record of the dates equipment was put in service or repaired provide the basis for establishing a maintenance schedule based on actual operating conditions. This record will allow routine-wear items to be identified and organized into groups for replacement at planned intervals. Scheduling periodic maintenance allows the operator the advantage of rinsing or clearing the concentrated fluoride chemical from the portion of the system scheduled for repair and gathering the tools and parts necessary for the job prior to beginning the work. This practice reduces the risk of unexpected operator contact with the chemical and provides customers with more continuous service. The state health department and/or state drinking water program should be contacted to obtain specific record-keeping requirements.

Laboratory analysis may be subject to additional state requirements on record keeping. Some common requirements are:

- Identification of the lab or facility at which the tests are performed.
- Identification of the equipment and materials used for testing. This includes a current and archived inventory of certificates of analysis for stock chemicals and reagents, preparation logs of the working solutions of those chemicals and reagents, and analytical instrumentation (including make, model, maintenance records, and calibration records).
- Identification of personnel involved in sample collection, receipt, preparation, and analysis.
- A COC that tracks the movement of a sample from collection to receipt by lab personnel performing the analytical testing.
- Detailed records of sample analyses, including any sample preparation (distillation) records, calibration and quality control testing associated with the sample, raw sample results, and data calculations (such as to account for analysis of a diluted sample).
- Provision of a sample location.

Examples of good practices to follow in a record-keeping system are listed below.

- Write legibly.
- Identify on each record the name of the person performing the documentation.
- Write in indelible ink.
- Keep the raw, original record (i.e., do not write on scratch paper and transcribe to a notebook or log without photocopying the original record and attaching it to the neater transcribed record).
- Cross out an incorrect record with a single line (i.e., do not erase or otherwise obliterate a written error). If a sample is reanalyzed, cross out the original entry with a single line and note "see below," where the reanalyzed record is written as if for a new, separate sample.
- For any records maintained electronically, ensure that they are backed up routinely or saved to a network drive.

It is also a good practice to maintain records of corrective actions for out-of-control events. The record should include the nature of the event, the steps taken to determine the cause(s), the cause(s) found, and the action(s) taken. Maintaining records of corrective action(s) should help minimize repeated re-occurrence of the same cause(s).

Operator Safety in Handling Fluoride Products

Fluoride improves health at the recommended level, but at higher acute exposure, there can be adverse effects. If an operator takes reasonable care and uses the correct personnel safety equipment, he or she can avoid the higher acute exposure possible when handling the concentrated fluoride products. Safety suggestions in AWWA Manual M3 *Safety Practices for Water Utilities* and fluoride product manufacturers' MSDS should be the basis for operator safety in water fluoridation.

The MSDS provided by the fluoride product supplier for its product is the reference for personal protection equipment (PPE) and safe handling of fluoride products; water facility personnel should refer to the recommended PPE. The recommended gear for fluoride is generally consistent with other additive products used at water facilities. In general, the consensus of manufacturers regarding recommended PPE is:

- Dry products: Long-sleeved overalls to keep the dust off clothes, leather gloves to protect skin from damage, goggles to protect eyes, and a National Institute for Occupational Safety and Health (NIOSH) dust mask.
- Liquid acid: Rubber apron, chemical resistant boots, long-sleeved gauntlet rubber gloves with the cuffs folded back, goggles for eye protection, and a face mask for protection from splashes.

Fluoride storage areas should be segregated from other additive storage areas due to incompatibility between various chemicals. All locations should be clearly and distinctly labeled to avoid accidents. All chemical deliveries should verify the correct location before unloading. There have been reports of incorrect deliveries resulting in accidents (such as sodium hydroxide being added to the storage tank of fluorosilicic acid resulting in a release of hydrogen and fluorine gases). Storage and handling areas should have appropriate ventilation and lighting for safe occupation.

The disposal of empty fluoride containers is usually a problem. The temptation to reuse fiber drums is difficult to overcome because the drums are convenient and sturdy. Paper bags are dusty and could cause a hazard if burned, and empty acid drums could still contain enough acid to cause personnel exposure. Empty fiber drums should be rinsed with water; paper bags should be folded and inserted into a plastic bag, then disposed of properly. Bags should not be shaken to purge the residual dust. Drums and other containers that have been rinsed should never be used where traces of fluoride could present a hazard. The solid waste division of the state environmental protection program should be consulted for correct disposal procedures. If possible, the fluoride storage area should be kept locked and not used for any other purpose. An eyewash station should be easily accessible.

Toxic Exposure

This section describes the difference between chronic and acute toxic exposure.

- *Chronic toxic exposure* occurs when exposure to large doses of fluoride takes place over a number of years. This type of exposure could occur as a result of repeated dust inhalation when feeder hoppers are being filled.
- *Acute toxic exposure* results from a single massive dose. Acute fluoride poisoning may result from ingestion, inhalation, or bodily contact with concentrated fluoride compounds.

The body's assimilation capacity for fluoride is limited. The body begins retaining fluoride at an exposure rate of about 1.5 mg/d, which would be the exposure from drinking water at the recommended fluoride level. It is estimated that chronic exposure may start at 0.20 to 0.35 mg/kg of body weight (for 175 lb [80 kg], that corresponds to 16 to 28 mg, or roughly 10 times the recommended fluoride level). With chronic toxic exposure, there may be an increased potential for skeletal fluorosis and increased bone brittleness.

Knowledge is limited concerning acute fluoride poisoning caused by ingestion or inhalation because it is very rare. Accidental ingestion is quite unlikely but could occur through contamination of food or drink either by mistaking the compound for sugar or salt or through carelessness in allowing areas where food is consumed to become contaminated by dust or spillage. The symptoms of acute poisoning by inhalation of a large quantity of dust or vapor include sharp biting pains in the nose followed by nasal discharge or nosebleed and possibly coughing or respiratory distress. Acid spill or splash may cause a tingling or burning sensation of the skin; if the eyes are involved, severe eye irritation may result.

Fluoride Overfeed

When a community fluoridates its drinking water, the possibility to overfeed fluoride can occur. Because of engineered safety considerations in facility installations, overfeeds are rare. Most overfeeds that do occur have no serious consequence, but all overfeeds should be taken seriously and corrected immediately. If an overfeed condition remains uncorrected, serious health consequences can result.

The CDC recommends the actions listed in Table 6-4 for fluoride overfeed instances. When a fluoride test result is near the limit of the analyzer's scale, the water sample must be diluted and retested to measure high fluoride levels accurately.

Maintenance

To ensure uninterrupted and consistent fluoride feed, proper maintenance of equipment is required. This includes maintaining not only the fluoride feeder but also all the appurtenances, feed lines, and laboratory testing equipment.

Fluoride Feeder. Some specific maintenance concerns that apply to fluoride feeders are cleaning, lubrication, access to spare parts, inspection, and recalibration.

Cleaning and lubrication details include the following:

- Like any mechanical device, fluoride feeders must be kept clean and lubricated in order to perform their function efficiently. A regular maintenance program will also minimize costly breakdowns and ensure long life for the equipment. Electric motors have a prescribed schedule for lubrication—the right type, amount, and frequency of lubrication are important. Gearboxes must be kept filled to the prescribed level with the proper lubricant, and all moving parts and unpainted metal surfaces should be kept clean and rust-free. If there are grease fittings, the proper grade, quantity, and frequency of lubrication should be observed.
- Many manufacturers have parts information available on their websites or listed in the provided operation and maintenance manuals.
- When fluoride feeders and related equipment are purchased, they should be accompanied by an instruction book and parts list. The instruction book will contain information on maintenance and repairs, and the parts list will enable the operator to select replacement parts when needed. If this information is lost, call or write the manufacturer or the manufacturer's representative for replacement if the manual is not found on the Internet. A list of spare parts should be kept on hand; having parts available can minimize the length of shutdowns as a result of equipment failure. In large water plants, an entire spare feeder should be available.

Inspection and Recalibration. Fluoride feeders will feed as intended only if the measuring mechanism is kept clean and operational. The following should be inspected regularly for signs of wear or damage: diaphragms or pistons of solution feeders; rolls, belts, disks, or screws of dry feeders; and associated mechanisms of both types. Repairs or replacements should be made before the critical equipment machine breaks down. Even with all parts in the best mechanical condition, fluoride delivery can be affected by leaks, spillage, buildup of precipitates from solutions, or accumulation of dry chemicals on or around measuring mechanisms. Occasional recalibration of the feeder can reveal evidence

Table 6-4 Recommended overfeed actions

Fluoride content in drinking water	Perform the following recommended actions
0.3 mg/L above control range to 2.0 mg/L	1. Leave the fluoridation system on. 2. Determine what has malfunctioned and repair it.
2.1 mg/L to 4.0 mg/L	1. Leave the fluoridation system on. 2. Determine what has malfunctioned and repair it. 3. Notify supervisor and report the incident to the appropriate state agencies.
4.1 mg/L to 10.0 mg/L	1. Determine what has malfunctioned and immediately try to repair it. 2. If the problem is not found and corrected quickly, then turn off the fluoridation system. 3. Notify supervisor and report the incident to the appropriate state agencies. 4. Take water samples at several points in the distribution system and test the fluoride content. Retest if results are still high. 5. Determine what has malfunctioned and repair it. Then, with supervisor's permission, restart the fluoridation system.
10.1 mg/L to 30 mg/L	1. Turn off the fluoridation system immediately. 2. Notify supervisor and report the incident immediately to the appropriate state agencies and follow their instructions. 3. Take water samples at several points in the distribution system and test the fluoride content. Retest if results are still high. Save part of each sample for the state lab to test. 4. Determine what has malfunctioned and repair it. Then, with supervisor's and the state's permission, restart the fluoridation system.
Exceeding 30 mg/L	1. Turn off the fluoridation system immediately. 2. Notify supervisor and report the incident immediately to the appropriate state agencies and follow their instructions. 3. Take water samples at several points in the distribution system and test the fluoride content. Retest if results are still high. Save part of each sample for the state lab to test. 4. Flush water mains to purge water exceeding 30 mg/L from the distribution system. 5. Public notification may be required by the state to prevent consumption of high levels of fluoridated water. 6. Determine what has malfunctioned and repair it. Then, with supervisor's and the state's permission, restart the fluoridation system.

Source: Engineering and Administrative Recommendations for Water Fluoridation, CDC

of potential malfunction. Feeders should be recalibrated at the frequency recommended by the equipment manufacturer.

Leaks. Leaks in and around the discharge line of a solution feeder can affect the quantity of solution delivered and result in low fluoride levels. Leaks are an annoyance and are always somewhat corrosive, ranging from the salt effect of sodium fluoride solutions to the acid corrosion of fluorosilicic acid. Even the smallest leak can result in damage to the feeder, appurtenances, or surroundings if left unattended. Leaks of strong solutions result in the formation of crystalline deposits, which, if allowed to build up, make subsequent cleaning difficult. A leak in the suction line of a solution feeder, though not immediately apparent, will adversely affect delivery and can eventually lead to air binding and cessation of the feed. Air binding can also be caused if fluoride solution is injected at the top of a main where air can collect.

Leaks in a dry feeder installation cause a dust problem, and if the feeding mechanism (for example, around the rollers of a volumetric feeder) leaks, the feed rate will be incorrect. In addition to an economic loss, the dust is a hazard to equipment and personnel. Leaks in equipment not related to fluoride feed can also present problems. For example, a water leak can result in dampness and subsequent caking of dry chemicals. A leak in a chlorine gas system can result in damage to feeders and associated equipment. Leaks in other dry feed equipment can result in dust contamination of the fluoridation installation.

Precipitates. When strong solutions are used, the possibility of precipitation buildup is usually present. In a solution feed system, precipitates in the feeder pumping chamber or on the check valves can affect delivery rate or stop the feeder entirely. Deposits in suction or feed lines can build up and stop flow. A coating of insoluble matter on a saturator bed can prevent water from percolating through. If the deposits are the result of water hardness, then softening the makeup water will eliminate the problem. If softening is impractical, frequent inspection and removal of the deposits is a necessity. Even when the water is soft, impurities in the chemical used and other mineral constituents in the water can build up to the point where small openings are clogged and the feed is impaired or stopped. Dissolving chambers and dry feed installations are vulnerable locations for precipitation or solids crusting. Frequent inspection and cleaning are the best solutions to this problem.

Tanks in which fluoride solutions are prepared invariably show precipitates of the insoluble impurities from the chemical used or of insoluble compounds formed by the reaction of the chemical with mineral constituents of the water. If a separate tank is used for fluoride solution preparation and the clear supernatant layer is then transferred to a day tank, then problems will be minimized but not necessarily completely eliminated. A regular schedule for cleaning a saturator should be established. The time interval between cleanings will depend on the amount of use and the accumulation of impurities in the saturator.

Storage Areas. Storage areas, although not necessarily a major factor in maintaining the accurate feed of fluoride, need to be kept clean and orderly. Bags of dry chemicals should be piled neatly on pallets. Whenever possible, whole bags should be emptied into hoppers. Partially emptied bags present a spillage hazard and are a nuisance to store. Empty fiber drums should be rinsed with water; paper bags should be folded and inserted into a plastic bag and then disposed of properly. Bags should not be shaken to purge the residual dust. Metal drums should be kept off the floor and tightly closed. The storage area for fluoride chemicals should be isolated from areas used to store other chemicals to avoid accidental mixing. All other types of materials, such as lubricating oil and cleaning equipment, should be kept out of the area. Both the stored chemicals and the general area should be kept free of dust, not only from the chemicals used in fluoridation, but also from other chemicals stored or used nearby.

Injection Facilities

A daily routine would include an inspection of the fluoride metering system, measurement and recording of the amount of chemical fed, collection of a fluoride sample at a designated state-approved sample point, and addition of the chemical to the feeder. Chemical usage and the amount of chemical added to the day tank or feeder should be maintained in a daily operator logbook along with the operator's name and date. In systems with multiple injection sites, all fluoride samples may be brought to the fluoride laboratory for testing the same day.

The operator should add fluoride chemical to the predetermined operating level to the saturator, dry feeder, or day tank as needed. Management should establish a maximum operating level for the chemical to be used for each injection site. This level should be high enough to allow feeding of the chemical for three days (long weekends) with empty head space to prevent overfilling the tank. In large plants, the day tank size might be reduced from three days capacity in consideration of shift changes or other considerations that may allow for a reduction of the day tank size. Operators should plan ahead to maintain the chemical in the hopper, saturator, or day tank for weekends, holidays, and higher than normal usage. Often, the removal of a source of supply for maintenance or repair puts a greater than normal demand on the operational wells or treatment plants.

If a system plans to discontinue fluoridation for a long time, the remaining fluoride in the day tank should be fed and the day tank and injection piping should be flushed with water. After flushing the suction line, it should be removed from the fluoride chemical supply or valved off from the chemical suction line for safety with lockout signage to help prevent any potential overfeeds (siphoning). Operators should keep the fluoride room door(s) closed.

REFERENCES

American Public Health Association, American Water Works Association, Water Environment Federation. 2012. *Standard Methods For the Examination of Water and Wastewater*, 22nd ed. Rice, E.W., R.B. Baird, A.D. Eaton, L.S. Clesceri, eds. Washington, DC: APHA, AWWA, WEF .

Brown, Richard. July 2014. Monitoring Fluoride, How Closely Do Utilities Match Target Versus Actual Levels? *AWWA Opflow.*

CDC. 1995. Engineering and Administrative Recommendations for Water Fluoridation. Atlanta, GA: CDC.

Chapter 7

Defluoridation and Managing Fluoride Levels

This chapter presents considerations on managing fluoride levels in distribution systems and reducing fluoride in sources that are excessively high.

HIGH FLUORIDE LEVELS

Both the USHHS and the Canadian Committee on Drinking Water (CDW) recommend a fluoride level of 0.7 mg/L. As fluoride levels increase to 2.0 mg/L, increasing beneficial protection from dental decay occurs, although the incremental increases in benefits are small. As the fluoride concentration continues to increase, however, the potential for fluorosis increases. Groundwater can leach fluoride from certain rock ores, resulting in elevated fluoride levels over what would be expected in surface waters.

The CDC has estimated that less than 4 percent of the US population is served by public water systems with naturally occurring fluoride at or above the level recommended for good oral health. The USEPA and CDC both have identified 2 mg/L as the threshold for increased potential for moderate to severe dental fluorosis, and the USEPA has set the SMCL, a cosmetic concern, at the 2-mg/L level. The CDC has identified based on state reporting to the CDC Water Fluoridation Reporting System that less than 0.5 percent of the US population is served by public water systems with naturally occurring fluoride at or above the 2 mg/L level. When a public water system has naturally occurring fluoride at that level, a water system must notify its customers once a year of the potential for staining of teeth in children under the age of eight years old. The customer notification must use specific language. The governing regulatory agency can provide the specific language for public notification in the event that a water system exceeds that level. In Canada, the CDW has set a maximum acceptable concentration (MAC) of 1.5 mg/L for protection from potential staining of teeth.

The USEPA has set the MCL for fluoride at 4 mg/L to protect against skeletal fluorosis, a rare condition with fewer than seven reported cases in the United States. The CDC has identified based on state reporting to the CDC Water Fluoridation Reporting System that only 16 states report having community water systems, serving less than 0.1 percent of the US population on public water systems, with fluoride exceeding the MCL of 4 mg/L. When naturally occurring fluoride exceeds 4 mg/L, the water system must annually notify its customers that fluoride exceeds the MCL; specific language must be used in the notification. The governing regulatory agency can provide the specific language for public notification in the event that a water system exceeds that level.

While most states have the same MCL for fluoride as the USEPA, some states have a lower state MCL. The governing regulatory agency can provide a water system detailed guidance on state-specific requirements concerning fluoride. Fluoride levels in excess of the MCL (or state-specific MCL) would be considered excessively high and require action. When fluoride levels exceed 2 mg/L, the water system operator should establish if this is true exceedance. As fluoride levels increase, the relative measurement error also increases, so a high measured level might be a laboratory artifact rather than a true measurement. If a water system suspects the reported fluoride level above 2 mg/L not to be respresentative of the true fluoride level based on historical reporting for that location or by calculation, then a repeat of the sample should be conducted to verify a high level.

STRATEGIES FOR MANAGING WELL FLUORIDE LEVELS

Rainwater and the resulting surface flows generally have a fluoride content of 0.1 to 0.2 mg/L. Water entering groundwater supplies has the potential for exposure to ores that may leach fluoride, although most rocks do not change the fluoride content of the water by leaching. If a well has a fluoride concentration greater than 4.0 mg/L, the water system can consider well management strategies including the following:

- Drilling a new well to obtain lower-fluoride water
- Connecting to an adjacent water system
- Specialized well construction and configurations to minimize high-fluoride water
- Using high-fluoride well supplies as an emergency supply
- Well blending to mix waters of different fluoride concentrations
- Defluoridation using either reverse osmosis or activated alumina

Often, the least expensive approach is to obtain an alternative source of water with lower fluoride content. This option can often be accomplished by drilling a new well or connecting to an adjacent or nearby water system if a source is within a reasonable distance.

High fluoride can originate in both igneous and sedimentary geological sources. In some cases, the fluoride source may be widespread, and all wells in proximity may share a high-fluoride characteristic. Several southwestern states, ranging from Texas to Arizona, have regional aquifers with high-fluoride characteristics from common regional ores. In other cases, the fluoride source may be from a particular geological or localized source. For example, Iowa has a discrete geographic source where an ancient meteorite or asteroid impact results in a small area with excessively high fluoride while adjacent areas do not display the same high-fluoride character. Similarly, numerous cases along the Eastern Seaboard exhibit one well with elevated fluoride and another well less than 500 feet distant with no fluoride content. Therefore, if high fluoride is encountered in the groundwater, the first investigation should be of regional and adjacent wells to ascertain how widespread the occurrence is. If all wells have consistently high fluoride levels, then presumably the regional ore deposits will influence all wells to be uniformly high. However, if other wells have varying fluoride levels, then localized geological influences may offer the potential to locate a lower-fluoride water supply.

Potentially, well modification can be used on a high-fluoride well to exclude the fluoride source. If fluoride is not uniformly distributed throughout the entire depth of the well, the fluoride content may be manageable through well modification. When designing and constructing new wells, it is important to study local well logs and design the casing and screen to avoid producing water with excess fluoride. Collecting multiple samples at different depths in the water column in the well may identify a particular range of depth with higher-fluoride water and other depths with lower-fluoride water. If fluoride varies by depth, construction grouting or well packers can be used to seal off locations found to produce high-fluoride water.

Although not common, locations exist where wells with different fluoride levels provide water to the distribution system at multiple locations. In a distributed supply system, zones of influence hydrodynamically associated with each well fluctuate daily and seasonally according to network characteristics and water demand. In this system, each neighborhood is principally supplied by a single well where water from different wells will not blend or mix, resulting in a fluoride concentration associated with an individual well. Two options are available when multiple wells exist with different fluoride levels. The lower-fluoride well can potentially be a primary supply if it is sufficient for the typical average annual demand, and the high-fluoride well could serve only as an emergency supply or as a seasonal supplemental supply. Another possibility is to blend a low-fluoride well with a high-fluoride well to obtain a mixture with an acceptable fluoride level. If wells with different fluoride levels are piped to a common location and mixed before the distribution system, the system fluoride levels can be equalized.

METHODS OF REMOVAL

Situations where a different source or well blending may not be feasible or possible require a water system to consider defluoridation or fluoride removal. Defluoridation alternatives tend to be expensive and operator intensive. Because fluoride is a dissolved ion, removal is technically challenging. The two major technologies for fluoride removal are activated alumina treatment process and reverse osmosis.

Activated Alumina

The use of activated alumina for defluoridation is rare due to operator requirements and necessary chemical adjustments. Activated alumina is a porous granular material possessing a large surface area per unit weight. Activated alumina is a by-product of aluminum manufacturing and is primarily aluminum oxide (Al_2O_3) activated by exposure to high temperature and caustic soda. The material is ground into a granular form that is available in several mesh sizes. The optimum mesh size for fluoride removal in water treatment systems is 28 through 48. Coarser mesh sizes have been used, but the adsorptive capacity is significantly lower due to the lower accessibility of adsorptive surfaces in larger grains. Finer mesh sizes have not been used (other than in laboratory bench-scale work) because the finer mesh material results in excessive pressure drop and backwash problems. Since physical and process characteristics may vary among activated alumina product manufacturers, the information provided herein is represented as typical. Therefore, the characteristics of a given activated alumina product should be verified by experience, special knowledge, or field pilot testing on the actual water to be treated.

Fluoride removal by means of activated alumina is an adsorption process in which fluoride ions in the source water are attracted to and held by activated alumina surfaces. The fluoride removal adsorption process is pH sensitive, and activated alumina attraction for the fluoride ion is the strongest at pH 5.5. Activated alumina fluoride capacity varies depending on the concentration of the fluoride in the raw water. For example, when the raw water fluoride concentration is 4.0 mg/L, the activated alumina fluoride capacity is 4,500 g/m^3 (2,000 grains/ft^3); and when the raw water fluoride concentration is 12 mg/L, the activated alumina capacity for fluoride is 9,000 g/m^3 (4,000 grains/ft^3).

As the pH of the water deviates either higher or lower from the optimum 5.5, the attractive forces (and the fluoride capacity) are reduced. At those pH levels, the previously adsorbed fluoride ions can be flushed from the activated alumina treatment media in a wastewater stream that requires disposal. Wastewater disposal must comply with regulatory agency rules. The removal of the adsorbed fluoride ions from the activated alumina is termed *regeneration*, the part of the treatment process in which the original capacity for fluoride removal is restored to the activated alumina treatment media for reuse in the fluoride removal treatment cycle. Activated alumina attraction for fluoride can be reduced at low and high pH levels, allowing regeneration. However, the acidic pH level required for regeneration is too aggressive, resulting in the dissolution of the activated alumina granular material. Therefore, activated alumina regeneration is accomplished with dilute caustic soda (or caustic potash) solution without loss of treatment media.

Arsenic (arsenate and arsenite) is also removed from water by the same activated alumina treatment process and similarly activated alumina is pH sensitive with an optimum pH for treatment of 5.5. In addition, if arsenic is present with fluoride in the water, it is preferentially adsorbed on the media with respect to fluoride. Therefore, arsenic removal will continue long after the activated alumina capacity for fluoride has been saturated. However, when arsenic is present in the water, its concentration is normally in the low parts-per-billion range while the fluoride is in the parts-per-million range. Except in cases where the level of the arsenic is very high, arsenic does not impact the activated alumina fluoride capacity. In those exceptions, a variation of the normal fluoride removal treatment process is implemented (Rubel and Williams 2004). Another factor to consider with activated alumina treatment is that arsenic clings to activated alumina surfaces more tenaciously than fluoride. Consequently, removal of arsenic from activated alumina surfaces requires a higher concentration regeneration caustic soda solution.

Other ions are also removed from water by the activated alumina adsorption process with optimum removal characteristics at pH 5.5. These ions can be either organic or inorganic and do compete with fluoride for adsorption sites. However, they are not common in drinking water. If they are present, they can be regenerated by the same process used for fluoride. Additionally, other ions, which are adsorbed by activated alumina at pH ranges other than that which is applicable to fluoride removal, are also regenerated with applicable pH adjustments. A typical example is silica (SiO_2), which is adsorbed in a pH range between 6 and 10. Silica is regenerated by means of an appropriate concentration caustic soda solution.

The common treatment process concept consists of two downflow treatment pressure vessels in a series in which fluoride removal treatment occurs in the first (lead) vessel. A process schematic flow diagram is shown in Figure 7-1 and illustrates the water flowing through both vessels. Fluoride removal takes place in a treatment band, which moves downward in the lead vessel as the media in initial contact with the water becomes saturated with fluoride. As the treatment band extends through the outlet of the first vessel, treatment is completed in the second (lag) vessel. Eventually, the treatment band is totally contained in the second vessel. At this time, the first vessel can be removed from service for regeneration while the second vessel is placed in the lead position. When the first vessel regeneration is complete, it is returned to service in the lag position. The use of a

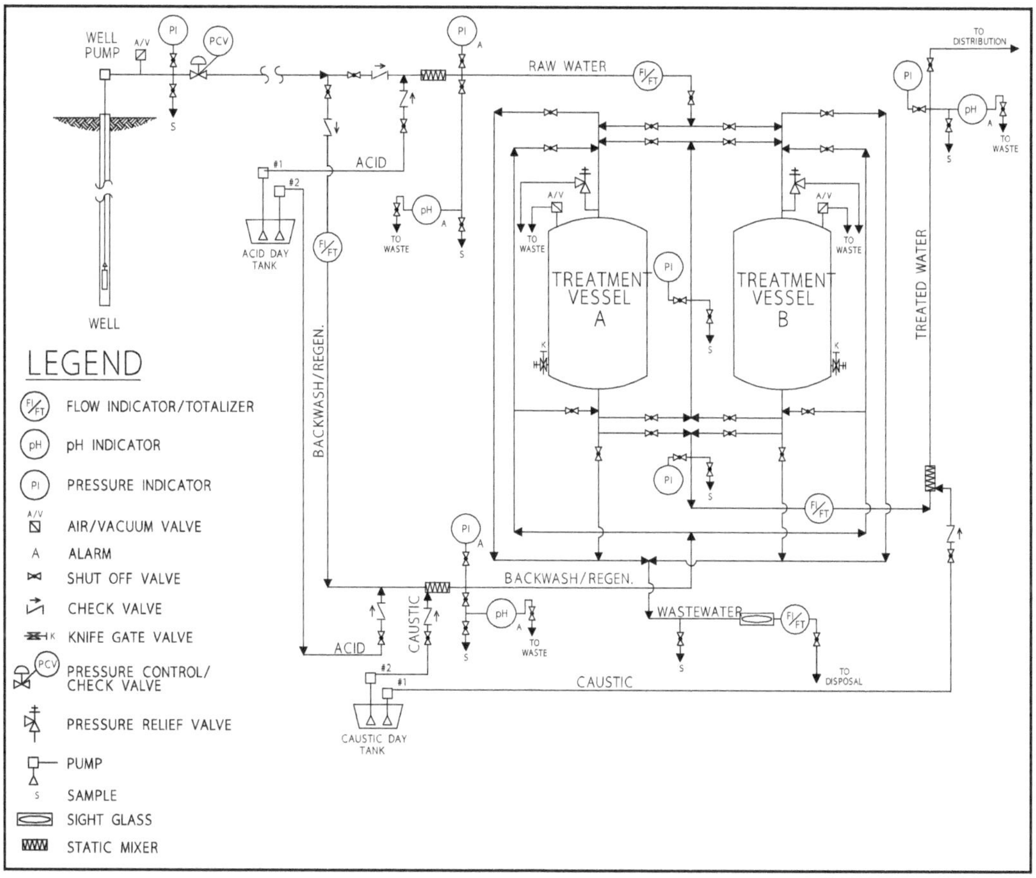

Figure 7-1 Activated alumina with pH adjustment fluoride removal water treatment plant

two-vessel train with the vessels alternating positions constitutes a continuous treatment operation that can perform indefinitely.

Treatment Plant Plan

The treatment plant described herein is for a central treatment for a well water supply. Components, other than the treatment system, are assumed to be provided separately and include a well, well pump, storage reservoir(s), pressurization subsystem(s), distribution subsystem, and disinfection subsystem. Because chlorine degrades activated alumina, it is applied after the defluoridation treatment. An activated alumina fluoride removal water treatment plant can be operated manually, semiautomatically, or automatically. A description of the features of each operational mode is beyond the scope of this chapter. An activated alumina fluoride removal water treatment plant based on the concept described here includes, but is not limited to, the following basic items:

1. Two identical treatment vessels are required.
   a. Each treatment unit is a vertical cylindrical pressure vessel. Skid-mounted supports are recommended to minimize foundation requirements.
   b. Each vessel is sized to contain the volume of activated alumina treatment media to treat the required flow rate. (Note: For flow rates that exceed the size available in standard pressure vessels, the feedwater should be proportionately divided into multiple dual-vessel trains that operate in parallel.) Each treatment

vessel will contain the volume of activated alumina treatment media that will provide a minimum empty bed contact time (EBCT) of five minutes. The depth of the activated alumina treatment media ranges between 3 and 6 feet. The vertical straight side height of the treatment vessel provides 50 percent activated alumina treatment media expansion during backwash plus a 6-in. freeboard. Therefore, a typical treatment vessel that employs an 8-ft straight side will include 5-ft activated alumina treatment media depth plus 2 ft, 6 in. for media expansion and 6 in. freeboard.

c. The treatment vessel will include internal piping, which uniformly distributes and collects the various process streams that flow through the activated alumina treatment media. Most importantly, the internal piping will include provisions to prevent activated alumina grains from exiting the treatment vessel.

d. All materials of construction must be suitable for exposure to acid and caustic in the pH range 2 to 13. The materials of construction or their protective linings must also have abrasion-resistant qualities. Both sulfuric acid and hydrochloric acid are used successfully for lowering pH, while both caustic soda and caustic potash are used successfully for raising pH. The sulfuric acid and caustic soda are usually less costly. Other pH-adjusting chemicals may be applicable, but each chemical requires evaluation of its characteristics.

2. Process water pipe, fitting, valve, and accessory materials shall also be suitable for exposure to pH in the range of 2 to 13.

3. Chemical bulk storage tanks for acid (usually H_2SO_4 66°Bé) and caustic (usually 50 percent NaOH) shall be sized to comply with the volume of the container by which the commodity is delivered to the treatment plant. For example, chemicals are the least costly when procured in bulk quantities, such as tank trucks or railroad tankcars. Additionally, storage tanks shall be designed with a minimum capacity sufficient for the full transportation vehicle load plus the maximum volume that can be consumed during a maximum time period required for delivery. Where fluoride-removal water treatment systems are not large enough to justify purchases of chemicals in bulk quantities (because of cost of bulk storage equipment, space requirements), the commodities are purchased and stored at a premium cost in small containers, such as drums and carboys. Regardless of the size or type of storage vessel, the materials of construction must be resistant to the chemicals to which they are exposed, and the storage containers must be set in secondary containment structures to prevent chemical leaks and spills from escaping into the environment.

4. Chemical feed pumps, feed systems, and so on, must be sized to accurately supply all feed rate requirements up to at least 110 percent of the maximum. Chemical feed equipment may be operated and controlled either manually or automatically. All chemical feed equipment and miscellaneous items exposed to treatment chemicals must be constructed of materials that are resistant to the chemicals for which they are employed. All feed equipment for treatment chemicals must be either double contained or shielded to provide complete operator safety. Emergency shower and eyewash equipment shall be provided at all locations where operators might be exposed to treatment chemicals.

5. Instruments and accessories are required to monitor or control pH, flow rate, total flow, pressure, and liquid level. As previously mentioned, the instruments and control components may be operated manually or automatically. Operation of an activated alumina fluoride-removal water treatment plant may be fully automated under the control of a programmable logic controller, or individual operational

features may be operated under the control of an individual controller. All instrument and accessory materials of construction must be recommended for all of the chemicals to which they might be exposed.

The treatment for removal of the fluoride takes place at pH 5.5, which is the optimal operating pH. The treatment may be continuous or intermittent. The pH of the treated water must be readjusted to a level that is acceptable for distribution (approximately 7.5). This can be accomplished by various methods, such as addition of caustic soda, blending with raw water, or stripping of free carbon dioxide by means of aeration equipment.

The initial startup of the treatment operation requires careful placement of the activated alumina in each treatment vessel. Prior to activated alumina application, each vessel should be half filled with water. This procedure separates the fine particles (powder) from the granular material, prevents cementing of the activated alumina particles, stratifies the particles by grain size, and protects the internal piping from impact loads. Prior to placing the treatment vessels into operation, the activated alumina treatment media should be backwashed until the backwash water is clear. At this point, the vessel should be drained, and the top ¼ in. of fine media should be carefully skimmed off the surface.

After completion of a treatment cycle (when the fluoride concentration of influent and effluent are the same), the treatment vessel is removed from service while treatment continues in the other vessel. The activated alumina must then go through a regeneration, which continuously employs raw water for all steps. The regeneration consists of the following steps:

1. Backwash the activated alumina treatment media at a flow rate of 7 to 9 gpm/ft^2 (depending on the density of activated alumina and water temperature) for a minimum of 10 minutes or until backwash water effluent is clear.
2. Drain treatment vessel.
3. Perform upflow regeneration using 15 gal of 1 percent NaOH solution per cubic foot of activated alumina treatment media. The flow rate shall be established to provide 30 minutes minimum for this step. For example, if the activated alumina treatment bed is 8 ft in diameter and 5 ft deep, the bed volume is 250 ft^3. Therefore, 3,750 gal of 1 percent NaOH will be used during the upflow regeneration step. Flow will take place at 125 gpm for 30 minutes. This equates to a flow rate of 2.5 gpm/ft^2 (50 ft^2 activated alumina treatment media surface area).
4. Perform upflow rinse. At completion of the upflow regeneration, the NaOH feed pump is turned off, but the water continues flushing fluoride from the activated alumina upflow at the same flow rate for 60 minutes.
5. At the completion of the upflow rinse, drain the treatment vessel to the top of the activated alumina treatment media.
6. Perform the downflow regeneration using the same procedure as the upflow regeneration (15 gallons of 1 percent NaOH per cubic foot of activated alumina treatment media for a minimum period of 30 minutes).
7. Perform the downflow rinse for 30 minutes at a flow rate that provides a minimum EBCT of 5 minutes. Using the example, activated alumina volume is 250 ft^3 (1,875 gal), which results in a maximum flow rate of 375 gpm (1,875 gal/5 min). If an EBCT of 7.5 minutes is used, the maximum flow rate would be 250 gpm.
8. Neutralization of the activated alumina treatment media with pH elevated to greater than 12 by the 1 percent NaOH regenerant solution takes place in three steps using the same raw water flow rate with pH adjustment by the feed of (sulfuric) acid.

a. The first step takes place at pH 2.5. This step continues until the pH of the effluent leaving the activated alumina treatment media drops to 8.5.
b. The second step takes place at pH 4.0. This step continues until the pH of the effluent leaving the activated alumina treatment media drops to 6.5.
c. In the third step which takes place at pH 5.5, the regenerated activated alumina treatment vessel is placedback into the treatment mode in the lag position.

The wastewater, which contains the fluoride removed from the activated alumina, along with the chemicals (acid and caustic) employed during the regeneration, may be disposed in one of several ways depending on local regulations and conditions. The most desirable disposal method is reuse for a beneficial purpose, such as irrigation or industrial application. Frequently, those opportunities are not available. Therefore, other disposal methods, such as evaporation ponds or sewer discharge must be evaluated. The cost of disposal in addition to potential environmental impact requires careful consideration. Fortunately, the volume of wastewater is small (approximately 3 percent) compared to the treated water produced.

Reverse Osmosis

Use of reverse osmosis (RO) treatment to manage fluoride levels has been the more common approach to defluoridation. When fluoride is high, there may be other undesirable characteristics of the water, and RO will provide a more comprehensive approach to finished water quality.

A complete description of RO membrane system design, equipment, installation, operation, and maintenance is the subject of AWWA Manual M46. Therefore, only a brief overview is presented herein.

A typical RO system consists of three subsystems: pretreatment, membrane process, and posttreatment. The feedwater is forced through the membranes under pressure, generating a product called *permeate*. The concentrate (reject stream), a waste stream that includes all of the rejected ions and a fraction of the feedwater, is also produced.

Pretreatment may include filtering, or depending on raw water quality, it can include acid addition and antiscalant to prevent precipitation of salts on the membranes. Pretreatment for the removal of membrane fouling compounds of the feedwater may also be required. The feedwater is filtered prior to entering the membrane stage. Polishing filtration is performed using low-micron cartridge filters.

The membrane system consists of a membrane array. Several elements are included in each array. The size of the elements can vary (from 4 to 16 inches in diameter) depending on the production requirements. The number of membrane stages used depends on the percent recovery needed and the fouling characteristics of the feedwater. A three stage array, for example, may produce a throughput of 85 percent to 90 percent. However, when high concentrations of silica exist in the feedwater, the throughput can be reduced, resulting in a larger volume of wastewater requiring disposal.

Posttreatment might include aeration to remove hydrogen sulfide, carbon dioxide, or other dissolved gases. This is followed by disinfection and corrosion control. The RO system can remove as much as 90 percent of the fluoride. Many RO plants are designed for split flow treatment and are particularly appropriate for fluoride removal installations. In those cases, part of the raw water bypasses treatment and blends with the treated water, resulting in a fluoride level in the blended finished water fluoride level meeting compliance requirements. Normally, these systems will aim for a blended finished water of 1.5 mg/L, a level below the SMCL but within the beneficial fluoride level. This strategy reduces the required size of the RO system and lowers treatment cost.

Proper disposal of concentrate from the RO treatment process involves the same considerations as the disposal of brine from the activated alumina process. In some states, it is regulated as an industrial waste, particularly if sufficient dilution is available. Permits for concentrate disposal must be obtained from the applicable regulatory agency.

POINT OF USE–POINT OF ENTRY

In certain systems, where central fluoride removal treatment is impractical, a few states may consider allowing a community to use point of use (POU)–point of entry (POE) home treatment units. The decision to implement POU home treatment units for compliance with a nonbiological MCL or regulation must be approved in advance by the state/provincial drinking water agency, and it must be available to all users. The Association of State Drinking Water Administrators reported in 2012 that approximately half of the states allow POU-POE for compliance in select cases but most states only approve that strategy for noncommunity systems. Thus, discussion with the local regulatory agency as to what is allowed in a specific state is important.

If POU-POE is used to address high-fluoride water, only the water intended for human consumption is treated. However, there is evidence that many consumers are dissatisfied with POU water and often bypass use of this water. POU water can be limited in supply from a concentrator tank, supplied as a trickle flow, and is warmer than water supplied directly from the distribution system. Therefore, it is not perceived by the consumer as aesthetically desirable when compared to free-flowing tap water with a cooler temperature. The POU approach requires careful definition of anticipated water supply zonal boundaries as discussed in previous sections of this chapter to ensure that diurnal/seasonal fluctuations in supply are addressed.

When POU-POE strategies are employed, it is important to have procedures in place that ensure consistent removal of fluoride at all residences. This responsibility falls upon the water utility or the operator of the water system, which includes utility ownership of the POU-POE devices and provision of those devices to each customer as a condition of approval by the regulatory agency. Additionally, the title of the property with a device must be encumbered with the rights to service and maintain a POU-POE device by the water system. Some residences will require a unique device, whereas other adjacent residences might be able to share a common device. A warning alert is also necessary if the device fails, and this requirement is often satisfied by including a total solids monitor.

Typical POU devices for fluoride removal use RO. According to NSF/ANSI Standard 58: Reverse Osmosis Drinking Water Treatment Systems, such devices are certified by the NSF to remove fluoride (and other constituents). The certification involves the use of a challenge water of 8.0 mg/L fluoride, and maximum permissible product water concentration of 1.5 mg/L. The devices have several owner-serviceable modules/filters, and maintaining performance is predicated on timely replacement of the membranes. The permitting of a POU-POE system will stipulate the frequency of monitoring, inspection, module replacement, and other maintenance.

REFERENCES

AWWA. 2007. M46, *Reverse Osmosis and Nanofiltration*. Denver, CO: AWWA.

Rubel F. Jr., and F. Williams. 2004. Pilot Study of Fluoride and Arsenic Removal From Potable Water. USEPA/600/2-80/100. Washington, DC: USEPA.

This page intentionally blank.

Chapter 8

Small Systems Considerations

This chapter focuses on the considerations of implementing fluoridation within small communities.

Small water systems are defined by the USEPA as those serving populations of fewer than 10,000 people. The beneficial impacts of adding fluoride to drinking water can be more significant for small water systems than for larger water systems. The decision to add supplemental fluoride can place financial, operational, and logistical constraints on small water systems that are proportionally larger in scale than for larger water systems. At the same time, the benefits can potentially be greater for small communities that may have limited access to dental care providers and may have a lower socioeconomic profile, which makes the benefits resulting from adjustment of fluoride the greatest. Whether the decision to add fluoride is mandated by the state, approved by the water system governing body, approved by a general election vote of the water system customers, or a combination of all three, special attention should be paid to ensure the fluoridation process is a sustainable match for the small water system.

A SMALL SYSTEM'S DECISION PROCESS: CONCEPTION TO CONSUMPTION

The decision process to add supplemental fluoride to drinking water in a small water system need not be a difficult undertaking. Contrary to the cultural viewpoint of the effectiveness and efficiency of a committee, the more individuals and groups involved in the process, the better the end product will be, assuming all parties work toward a common goal and not against one another. This can be especially important when dealing with a situation in which different opinions arise and a consensus must be achieved. A good facilitator may be essential in improving the potential to achieve a consensus.

The process often begins with the water system governing body, the water system/public works staff, and the water customers having discussions about the advantages and disadvantages of supplemental fluoride in drinking water. These discussions or meetings need to be sufficient to vet the array of opinions and identify public health benefits as well as public health concerns. An unsupportive public and an uninformed governing body can place constraints and burdens on the overall success of any project that will ultimately lead to failure, especially a project concerning supplemental fluoride. The state oral health director, a state dental director with the health department, or a regional public health board should be able to provide professional assistance on the costs, benefits, and community impacts. A balanced community discussion with a facilitator can be effective in forging collaborative agreements and understandings within the community. Since most small community water systems do not have public relations staff to facilitate these kinds of meetings, water system officials may want to seek out the advice and assistance of consultants or public agencies such as a state university Extension Service, the Rural Community Assistance Partnership (RCAP 2009), engineering consultants, or other entities as appropriate.

Once the project has moved past the preliminary approval and authorization stage, the small water system will transition to the design stage of the project. At this point, many public works directors, governing bodies, and water superintendents (the water system stakeholders) abdicate their role to design engineers to plan, design, and manage the construction/installation of the system. In some states, the Rural Water Association may provide limited professional services assistance on plant improvements.

Very few small water systems have their own city or district engineer on staff. Fluoridation is a specialty chemical application, and it may be appropriate to identify a firm that has actually designed a fluoridation system and has appropriate technical experience. It may also be appropriate to undertake a request for proposal/qualification process for this specific project. Either way, the water system stakeholders must work in partnership with their engineering service provider to ensure that the design of the fluoride feed system fits the needs of the community. The engineering plans and specifications for the system should apply specifically to the water system and the project that is being undertaken; the capital costs estimated by the engineer should be a reflection of these plans. The design should be reasonable and affordable to the community, and the actual construction of the fluoride system must be completed correctly and carefully with the materials and products specified in the plans.

CONSIDERATIONS FOR SIMPLIFIED SYSTEMS

The design and planning process for adjusting fluoride in drinking water is an important step for any size water system, but it can be challenging for some small water systems that may not have engineering staff or water quality technical staff to lead and guide the process. Many small water systems rely on consultants, the state Rural Water Association technical staff, and/or suppliers to manage the planning, design, and installation process for capital improvements. Another excellent resource is the AWWA section, which can often offer operator-to-operator assistance by pairing a smaller system with other communities that have already implemented water fluoridation. This pairing allows the system that is new to fluoridation to benefit from the experience of a system that is already fluoridating. While this process can be very successful, small water system officials must be diligent to ensure the design and operation of the fluoride system is compatible with the conditions and constraints of the existing water system, including the level of operator certification that may be required.

Choosing the Right Chemical

Those implementing the water fluoridation system must evaluate the forms of fluoride and choose the best fit for the system. Key considerations include the cost of the chemical, the chemical handling methods available to the small water system operator, and the amount of safety precautions and training needed for the operator to maintain a safe and reliable system. This review process should include the operations, management, supply, and consulting resources that are involved in ensuring the water system is operationally compliant. This process will be different for each utility and should account for regional considerations such as chemical availability, shipping costs, and the experience of other small utilities in the area.

Dry Sodium Fluoride. As stated in preceding chapters in this manual, dry sodium fluoride is the least expensive capital cost and relatively easy to maintain in solution. It is therefore often a preferred fluoride type selected by small water systems. Sodium fluoride is typically added using a saturator system that is particularly suitable for use by small systems due to its simple mechanics and ease in operation. Sodium fluoride is generally more expensive compared to other types of fluoride products when bulk quantities are purchased, but smaller systems purchase smaller quantities of fluoride products. When purchasing small quantities of products, the cost of sodium fluoride can be favorable in comparison.

Sodium Fluorosilicate. Similar to sodium fluoride, sodium fluorosilicate is a dry crystalline form of fluoride that is handled in bags or drums. However, sodium fluorosilicate requires more complex and expensive feeder equipment as it cannot be used in saturators. The feeder requires a hopper feed, regular calibration, and a mixing tank with a resulting unsaturated solution pH more acidic (3 to 4) than that of sodium fluoride (7 to 8). Sodium fluorosilicate is most commonly used for medium-size communities and tends to be the least expensive fluoride, particularly in small volumes.

Fluorosilicic Acid. Fluorosilicic acid can be a simple installation when used at a large water facility, but it does not scale down well and therefore smaller systems must include everything that is involved in a large system. This can render the acid feed to be the most complex and expensive to implement for a small system. Fluorosilicic acid is a strong-acid with a pH close to 1, and personnel protection during product handling is a higher concern than for the crystalline products. However, in high-hardness groundwaters, there can be a benefit in using fluorosilicic acid as the hardness can be an operational interference in dissolution of sodium fluoride. In bulk, the fluorosilicic acid product can be purchased directly from manufacturers, but for the small quantities that small systems would use, it is repackaged in carboys or drums through local distributors or product representatives. This repackaging in smaller delivery containers normally results in fluorosilicic acid being the most expensive fluoride product compared to alternative sodium fluoride or sodium fluorosilicate products.

Facility Considerations

At small water systems, the key to effective operations of fluoride systems is to maintain a safe and regular practice. The operation of the fluoride system needs to match the needs, goals, and abilities of the operator and the water system. The safety and protection of the operator is a key consideration in small system operation because, in many circumstances, an operator of a small water system is working alone, either in the field or in the treatment facility. Whatever operational practice is employed by a small water system, it must be accomplished on a routine or regular basis. The water system's ability to operate and sustainably maintain the fluoride system is as important as the ability to finance the capital cost of the equipment. It is for these reasons that a sodium fluoride system using a saturator is typically an excellent match for a small system.

When considering options, water system managers, governing boards, operators, and engineers can arrive at the false conclusion that bigger, more complex treatment and more moving parts equates to a better system. Throughout the country, there exist examples in small communities where technology has been applied incorrectly to solve a problem or remedy a condition. At the same time, simple should not imply a bare-bones or no-frills installation. Fluoridation is no exception.

The proper application of a fluoride feed system must take into consideration the current and anticipated conditions of the system itself and the operators who will be charged to manage and maintain the system. For example, to operate a fluoride system using dry fluoride, the operator must be capable of physically handling bulk bags of powdered chemical with the appropriate safety equipment (such as eye goggles, apron, respirator or dust mask, and gloves), and the treatment facility must have ventilation and washing/rinsing provisions built into the chemical feed area. Similarly, operators of liquid fluoride feed systems must be capable of easily moving barrels and carboys of chemicals while using appropriate safety equipment. The carboys and barrels as well as the day tank and feed equipment must be housed in areas where proper chemical spill containment provisions have been designed and constructed. Containment provisions and other safety and operational features can be easily overlooked in fluoride feed system projects for small systems. This is due to limited capital budgets, lack of experience with the installation of fluoride systems, and an underestimation of the importance of the provisions themselves.

Operational Practices

Operators of small water systems are generally not specialists concentrating on a small, defined set of tasks. Rather, these operators must possess expertise in a wide array of specialty areas because they bear the responsibility for managing and maintaining many systems such as the water, wastewater, stormwater, and transportation systems in small communities. Furthermore, at any given time, the "normal routine" of any one of those systems (water line break, road failure, wastewater treatment plant issue, clogged storm drains, for example) can impact all of the remaining system operation routines simply because there are not enough hours in the day to accomplish all of the daily, weekly, or monthly routine tasks. Therefore, establishing a routine set of operating tasks through the creation of a working set of standard operating procedures (SOPs) is critical. For a small system operator, the SOPs must realistically match the level of service the water system has committed to its customers, and it must match the level of staff availability within the water system.

The SOPs to operate and maintain fluoride feed equipment should describe what is required to (1) deliver fluoridated drinking water to customers safely, (2) maintain the functionality and integrity of the equipment, and (3) ensure those tasks can be completed with the staff available to the water system. It is important to take care that the SOPs do not simply describe what the water system could or should do with a fully staffed workforce. Many excellent asset management software resources exist to aid small systems in the creation of SOPs, maintenance schedules, and guidelines for typical staff time needed, and to track conditions of equipment. Water systems must also consider sampling and laboratory verification as part of the fluoride system SOPs to comply with the monitoring requirements of operating a chemical feed system. Operators must add daily fluoride tests and periodic certified laboratory sampling and sample delivery to their own list of responsibilities.

Another operational strategy used by some small systems, particularly concerning fluoride, is to identify a contract operator to assist in the fluoridation system operation. It may be appropriate to have the operations of the fluoridation system contracted to either an adjacent utility or a contract operator that has experience with fluoridation. In some locations, the chemical supplier has been known to provide enhanced service by making a daily visit to the fluoridation facility as part of its services.

REFERENCES

Massachusetts Department of Environmental Protection: Bureau of Resource Protection – Drinking Water Program. June 2010. *Water Supply Facility Checklist for Fluoridation.* http://www.mass.gov/eea/docs/dep/water/approvals/year-thru-alpha/m-thru-s/nafckl.pdf.

Massachusetts Department of Environmental Protection. *Preventive Maintenance Card File for Small Public Water Systems Using Ground Water.* http://www.mass.gov/eea/docs/dep/water/drinking/alpha/i-thru-z/sspmlcs.pdf.

Rural Community Assistance Partnership. November 2009. RCAP's approach to how we serve rural communities. http://www.rcap.org/node/75.

US Environmental Protection Agency. Information on Check Up Program for Small Systems (CUPSS) Asset Management Tool. http://water.epa.gov/dwcapacity/information-check-program-small-systems-cupss-asset-management-tool.

West Virginia Department of Health and Human Resources. *Class 1 Water Operator Course Manual – Fluoride Feed Systems*: 104–115. http://www.wvdhhr.org/oehs/eed/swap/training&certification/drinking_water/documents/ClassIManual.pdf.

This page intentionally blank.

Chapter 9

Community Outreach and Communication

This chapter focuses on community outreach and communication. An essential part of working with the community is providing excellent customer service, including answering questions that involve community members' health. Although communication is a topic that extends beyond fluoridation, fluoride issues are of interest to communities and their communications deserve special consideration.

CUSTOMER SERVICE AND COMMUNICATION

The nature of the drinking water industry as the "silent industry" may challenge water professionals if issues or concerns regarding community water fluoridation occur. Many members of a community take an interest in their health, and they want reliable information on the quality of their drinking water. Community education and communication in this regard are important aspects of being a public utility.

Questions about fluoride fall into the category of health effects. Is fluoride healthy or detrimental to health? Systematic scientific studies, such as the 2006 NRC of the NAS, have consistently agreed that water fluoridation at the appropriate level is not unhealthy, but rather beneficial for dental health. Numerous governmental health entities, professional medical associations, and allied institutional associations recommend the adjustment of fluoride to beneficial levels. In spite of the evidence, some members of the public continue to believe that fluoride should not be added to drinking water and have concerns about the practice. Drinking water industry professionals generally understand and agree that fluoridation of drinking water affords significant oral health benefits and does so in a safe and cost-effective manner. They are often the initial contact and respond as best they can to public inquiries of concerns about or objections to water fluoridation.

FLUORIDE COMMUNICATION STRATEGIES

Multiple strategies can be employed as part of a communications plan to help address concerns about fluoridation, including relationship building, strong ongoing community outreach, and effective media relations.

Building Relationships

The principal concerns about fluoride in drinking water relate to perceived adverse health effects. Many water utility staff members are not familiar with the relevant scientific health facts and, consequently, they may not be the best responders to this concern. Concerns may be best addressed by establishing active partnerships between dental and public health professionals who may be more knowledgeable about the science and health facts. A key resource may be the state dental director, normally an official in the state department of health. Some states will also have regional dental directors that could also provide assistance in addressing questions.

Resources from respected and authoritative sources on water fluoridation are available to provide information when responding to public questions as well. The CDC includes information on water fluoridation at www.cdc.gov/fluoridation. The American Dental Organization offers an excellent publication entitled "Fluoridation Facts" at www.ada.org; the information is periodically updated to address common questions that the public has about water fluoridation.

Community Outreach

While utilities can request state or local health agencies, elected officials, and appointed governmental advisory commission members to take an active role in community outreach and educational initiatives, the utility itself may be the most knowledgeable source of information to its customers. The utility should know all the facts on the safety and reliability of the equipment used for fluoride addition, the frequency of the monitoring the process to ensure customers are receiving the appropriate levels of fluoride in their water at the tap, and the source and required specifications of the fluoride chemical that will be added to the water.

Personnel within the utility's customer service and public affairs divisions should be prepared to answer basic questions about the practice of fluoridation, and should have contact information for a technical representative to whom customer inquiries may be directed for more information. A good answer to any health-related question should be: "You should discuss this with your doctor, dentist, or other personal health care provider." Customers with health questions may also be referred to local, regional, or state health agencies.

Additionally, utilities must provide adequate notification to the public prior to initiating fluoridation. Not only does early notification prevent customers from perceiving that a change was made without informing them, but such notification may be required by the authority regulating the utility's operational permit for fluoridation. Informational pamphlets, newspaper articles, direct mailings, paid advertisements, or radio and TV announcements are examples of the forms of communication typically considered acceptable. Direct notification of medical and dental professionals in the affected area can be beneficial, and may be required; engaging medical professionals gives them the opportunity to prepare for consultation with potentially affected patients if necessary.

Ongoing communication will be necessary with the community even after fluoridation has been implemented. An effective tool for this communication is the annual Consumer Confidence Report, which is required by the USEPA.

Media Relations

When a new research study is released or a new fluoridation strategy is launched, it hits the news. Whenever water fluoridation is in the news, public interest is sparked and water utilities may receive increased customer inquiries related to water fluoridation. Proactive media relations and education in advance of a fluoridation project will help mitigate some of the media coverage while helping disseminate the utility's messaging about fluoridation. Additionally, as mentioned previously, the appropriate staff should be prepared to respond to media and customer inquiries that the utility receives.

This page intentionally blank.

Appendix A

Materials Compatibility Lists

Appendix A presents three materials compatibility lists, one each for fluorosilicic acid (FSA), sodium fluoride (NaF), and sodium fluorosilicate (SFS). Many materials compatibility lists are produced by specific chemical companies or equipment manufacturers. Often, the conditions under which the exposure tests were run are not specified, which causes difficulties in interpretation for a specific application. In other cases, there are conflicts as to materials compatibility between lists from different sources.

The lists presented are experiential, and they derive from the operation of water fluoridation systems. The materials identified are known to provide good service under continental US conditions in water fluoridation systems using the fluoridation chemicals identified in the titles. As these lists are not exhaustive, AWWA requests that member utilities with experiential data from the operation of water fluoridation systems provide that information, so that additional materials that have proven useful over long periods of time can be added.

A short list of references to compilations of materials compatibility data is provided for those who want to possess such reference tools.

REFERENCES

De Renzo, J.D., ed. 1985. *Corrosion Resistant Materials Handbook,* 4th ed. Noyes Data Corp.

Lopez, M.T., and W. Sleeper. 2008. Corrosion of Metal Alloys in Water Fluoridation Systems, *CA-NV.* AWWA Spring Conference.

Schweitzer, P.A. 2004. *Corrosion Resistance Tables (Parts A-D),* 5th ed. Marcel Dekker.

MATERIALS COMPATIBLE WITH FSA

Tanks

Cross-linked polyethylene
HD linear polyethylene
Kynar-lined steel (exceptional in tank trucks)
Fiberglass – vinyl ester resin: Hetron 922 (Ashland) or Derakane 411 (Dow) with double synthetic veil and MEK-P cure
Rubber-lined steel (typical for railcars, tank trucks)
Composite: X-linked and HD linear polyethylene (HD linear in contact with FSA)

Piping (Light Blue with Red Bands)

PVC
CPVC

Solution Injectors (Neat FSA)

CPVC
Monel
Illium PD
Hastelloy C (when diluted 100–200:1)

Pump Seals

Teflon
Viton
Plain carbon (with no rubber parts)

Elastomers

Neoprene
Hypalon
Viton
EPDM

Concrete

Epoxy coating: 2 layers of epoxy + 1 layer of urethane
Furan-based cements

DO NOT USE

316 stainless steel
Copper tube
High and low zinc brass
Phosphor bronze
Cupronickel
Titanium
Buna-N rubber seals
Fluorosilicone

QUESTIONABLE

304 stainless steel
Hastelloy C
Illium G

AWWA acknowledges the valuable experiential information provided by: the Los Angeles Department of Water and Power and Metropolitan Water District of Southern California in the development of this materials compatibility list.

MATERIALS COMPATIBLE WITH NAF

Tanks or Saturators

Polyethylene (medium density)

Pump Seals

Teflon
Aflas

Tubing or Piping

Medium-density polyethylene
PVC
PVDF

Injectors

PVC
PVDF
20Cb-3 stainless (Carpenter Technologies)

DO NOT USE

Copper tube in contact with NaF solid or solution
Brass materials in contact with NaF solid or solution
Ceramic balls in pump check valves; replace with Teflon

AWWA acknowledges the valuable experiential information provided by: Northeast Pump & Instrument, Lunenburg, Mass., in the development of this materials compatibility list.

MATERIALS COMPATIBLE WITH SFS

Storage

Packages shall be stored off the floor on pallets or other support in a dry, low-humidity structure that is protected against water intrusion.
Packages shall be contained in multilayer bags with a moisture barrier; supersacks shall be lined with a moisture barrier.

Concrete (Floor of Receiving, Storage, and Handling Buildings)

Clear, acid-resistant coating (example particulars under investigation)

Tanks

Steel or stainless steel; lined with a flexible, elastomeric 100% solids epoxy with similar chemical resistance and cured system physical properties to Duromar DF-3710 at a total dry film thickness of 30–36 mils

Dry Chemical Feeders

316 stainless steel
Hastelloy C

Piping

PVC
316 stainless steel

Solution Injectors

PVC (Schedule 80)
316 stainless steel (for higher-pressure applications)

Pump Seals

Water seal with graphite packing

DO NOT USE

Mild steel
Unlined pumps
Noncorrosive resistant materials

AWWA acknowledges the valuable experiential information provided by: KC Industries, Mulberry, Fla., in the development of this materials compatibility list.

Appendix B

Example Calculations

PRELIMINARY CALCULATIONS

The first step in planning is to determine the optimal fluoride concentration and dose. Subsequently, based on the system pumping rates, chemical purity, and available fluoride ions, the daily quantity of chemicals needed can be determined. The quantity largely determines the chemical that should be used, which in turn determines the types of feeders that can be selected. Example calculations will be presented after each factor has been explained.

Recommended Fluoride Concentration

The recommended fluoride concentration recommended by the US Health Service for a community drinking water system is 0.7 mg/L.

Fluoride Dose

The dose is the amount of fluoride chemical needed to obtain the optimal fluoride concentration in drinking water. The dose, expressed as milligrams per liter (mg/L) or parts per million (ppm), is obtained by subtracting the naturally occurring fluoride concentration from the desired fluoride concentration. For example, if the desired fluoride concentration is 0.72 mg/L and the natural fluoride concentration is 0.2 mg/L, the dose is 0.5 mg/L (0.7 – 0.2 = 0.5).

System Pumping Rates

The maximum pumping rate, also called plant capacity, refers to the maximum flow of water that can be produced by a facililty. The capacity may be measured in gallons per minute (gpm), millions of gallons per day (mgd), cubic meters per second (m^3/s), or megaliters per day (MLD). The fluoride feed rate depends on the flow of water through the system. The maximum design feed rate will be based on the maximum pumping capacity, even if the facility does not currently reach that maximum rate. Similarly, the amount of

Table B-1 Typical values of purity and available fluoride ion (AFI)

Chemical	Purity, %	AFI	Available F, %
Sodium fluoride	98	0.452	44
Sodium fluorosilicate	98.5	0.607	60
Hydrofluorosilicic Acid	23	0.792	18

chemical needed for an average day is based on the average daily production of water. The average daily production of water is a quantity, such as 2 million gallons (7.57 megaliters) rather than a rate, such as 500 gpm (2.73 MLD). A facility may have a capacity of 2 mgd (7.57 MLD), but if it operates at that rate for an average of only 12 hours a day, then the average daily production is 1 million gallons (3.79 megaliters). However, every plant has daily and seasonal variations, and the actual feed rate is time dependent, varying with the flow of water through the system.

Chemical Purity and Available Fluoride Ion Concentration

Only a portion of the bulk quantity of chemical contributes to the optimal fluoride concentration. Chemicals as delivered are not 100 percent pure, and the fluoride ions may be bound to other elements. Table B-1 lists examples of purity and available fluoride ion (AFI). The percent available fluoride is the purity multiplied by the AFI.

Chemical Quantity

The daily quantity of chemical used can be determined from the daily production, the desired fluoride dose, the chemical purity, and the AFI. The quantity of chemical can be calculated by the following equation:

$$\text{Chemical usage (lb/day)} = \frac{\text{V (mgd)} \times \text{D (mg/L)} \times 8.34 \text{ (lb/gal)}}{\text{AFI} \times \text{P}}$$

$$\text{Chemical usage (kg/day)} = \frac{\text{V (MLD)} \times \text{D (mg/L)}}{\text{AFI} \times \text{P}} \qquad \text{(B-1)}$$

Where:

V = daily production
D = dose, in mg/L
AFI = available fluoride ion, decimal fraction
P = chemical purity, decimal fraction

Chemical Feed Rate

Calculation of the chemical feed rate is necessary for proper adjustment of the fluoride feeder. Dry chemical feed rates are calculated the same way as chemical quantity, except utilizing the plant flow rate or pumping rate instead of daily production. Chemical solution feed rates require the plant flow or pumping rate, desired dose, solution density, and solution strength.

The additive feed rate for dry salt product, such as sodium fluoride or sodium fluorosilicate, can be calculated using the following equation:

$$\text{Dry chemical feed rate (lb/min)} = \frac{\text{PR (mgd)} \times \text{D (mg/L)} \times 8.34\ \text{(lb/gal)}}{\text{AFI} \times \text{P} \times 1{,}440\ \text{(min/day)}}$$

$$\text{Dry chemical feed rate (kg/min)} = \frac{\text{PR (MLD)} \times \text{D (mg/L)}}{\text{AFI} \times \text{P} \times 1{,}440\ \text{(min/day)}} \qquad \text{(B-2)}$$

Where:

PR = plant flow rate or pumping rate
D = dose
AFI = available fluoride ion, decimal fraction
P = chemical purity, decimal fraction

For chemical solution feed rate, for a fluoride solution generated from dry chemical, the following equation may be used:

$$\text{Chemical solution feed rate (gal/hr)} = \frac{\text{Dry chemical feed rate (lb/min)}}{\text{Solution density (lb/gal)} \times \text{Solution strength}} \times 60\ \text{(min/hr)}$$

$$\text{Chemical solution feed rate (L/hr)} = \frac{\text{Dry chemical feed rate kg/min)}}{\text{Solution density (kg/m}^3\text{)} \times \text{Solution strength}} \times 1{,}000\ \text{(L/m}^3\text{)} \times 60\ \text{(min/hr)} \qquad \text{(B-3)}$$

Where:

Solution strength – percent weight of chemical in the solution, converted to decimal fraction

For chemical solution feed rate, for a commercial fluoride solution, the following equation may be used:

$$\text{Chemical solution feed rate (gal/hr)} = \frac{\text{PR (mgd)} \times \text{D (mg/L)} \times 8.34\ \text{(lb/gal)}}{\text{Solution density (lb/gal)} \times \text{Solution strength} \times \text{AFI} \times 24\ \text{(hr/day)}}$$

$$\text{Chemical solution feed rate (L/hr)} = \frac{\text{PR (MLD)} \times \text{D (mg/L)}}{\text{Solution density (kg/m}^3\text{)} \times \text{Solution strength} \times \text{AFI} \times 24\ \text{(hr/day)}} \times 1{,}000\ \text{(L/m}^3\text{)} \qquad \text{(B-4)}$$

Where:

PR = plant flow rate or pumping rate
D = dose in mg/L

Solution strength = percent weight of fluoride in the solution, converted to decimal fraction

AFI = available fluoride ion, decimal fraction

Note that for commercial solutions, the solution strength indicated by the supplier is typically the weight percent of pure chemical. The supplier should furnish an analysis that provides the actual fluoride content of the solution. As noted in Table B-1, the AFI for a 23 percent hydrofluorosilicic acid solution is 0.792, so the actual fluoride content in the solution is 18.2 percent.

Many states require records be maintained that include the amount of chemical used. They also require the theoretical concentration of chemical in water be calculated. This calculated concentration is a safety precaution to verify that an overfeed did not occur. Furthermore, it aids in solving troubleshooting problems. If the calculated concentration is significantly higher or lower than the measured concentration, steps should be taken to determine the reason for the discrepancy.

The following equation can be used to determine the calculated concentration (in milligrams per liter):

$$\text{Calculated concentration (mg/L)} = \frac{\text{(pounds of chemical)} \times \text{AFI} \times \text{P}}{\text{(million gallons of water treated)} \times 8.34 \text{ (lb/gal)}}$$

$$\text{Calculated concentration (mg/L)} = \frac{\text{(kilograms of chemical)} \times \text{AFI} \times \text{P}}{\text{(million liters of water treated)}} \tag{B-5}$$

The numerator of the equation represents the mass of fluoride ion added to the water, and the denominator represents the volume of water treated.

For facilities that utilize liquid chemical, the following equation can be used to determine the calculated concentration based on the volume of chemical used to treat a volume of water:

$$\text{Calculated concentration (mg/L)} = \frac{\text{(gallons of chemical)} \times \text{solution density (lb/gal)} \times \text{solution strength} \times \text{AFI}}{\text{(million gallons of water treated)} \times 8.34 \text{ (lb/gal)}}$$

$$\text{Calculated concentration (mg/L)} = \frac{\text{(liters of chemical)} \times \text{solution density (kg/m}^3\text{)} \times \text{solution strength} \times \text{AFI}}{\text{(million liters of water treated)} \times 1{,}000 \text{ (L/m}^3\text{)}} \tag{B-6}$$

A sodium fluoride feed system is unique because the strength of the solution formed in the saturator is always 18,000 mg/L. This is due to the fact that sodium fluoride solubility is practically constant at 4.0 g/100 mL of water at the temperatures generally encountered in water treatment. Each liter of solution contains 18,000 mg of fluoride ion (40,000 mg/L times the percent available fluoride [45 percent] equals 18,000 mg/L). The constant strength simplifies calculations, as the plant capacity (or pumping rate) and desired dose are all that are required. Thus, the chemical feed rate is as follows:

$$\text{Chemical feed rate} = \frac{\text{Plant capacity} \times \text{dose (mg/L)}}{18{,}000 \text{ (mg/L)}} \tag{B-7}$$

The feed rate will have the same units as the plant capacity. If the plant capacity is in gallons per minute, the feed rate will also be in gallons per minute.

Eq B-7 can be re-arranged to determine the calculated concentration based on the volume of saturated sodium fluoride solution used to treat a volume of water:

$$\text{Calculated concentration (mg/L)} = \frac{\text{Volume of solution} \times 18{,}000 \text{ (mg/L)}}{\text{Volume of water treated}} \tag{B-8}$$

EXAMPLE FEED RATE CALCULATIONS

Sodium Fluorosilicate

Example 1. The optimal fluoride concentration has been determined to be 0.7 mg/L, and the natural fluoride concentration is 0.2 mg/L. The chemical supplier has indicated that the commercial purity of the sodium fluorosilicate supplied is 98 percent. The solution density in the dissolving chamber is 8.34 lb/gal and the solution strength is 0.2 percent as sodium fluorosilicate. If the plant rate is 2,000 gpm, determine the dry chemical feed rate in pounds per minute and the chemical solution feed rate in gallons per hour.

Given:

Optimal fluoride concentration = 0.7 mg/L
Natural fluoride concentration = 0.2 mg/L
Chemical purity = 98 percent
Pumping rate = 2,000 gpm = 2.88 mgd
Chemical solution density = 8.34 lb/ft^3
Chemical solution strength = 0.2%

Solution:

a. Determine fluoride dose:

$$\begin{aligned} \text{Dose (mg/L)} &= \text{optimal fluoride level (mg/L)} - \text{natural fluoride level (mg/L)} \\ &= 0.7 \text{ mg/L} - 0.2 \text{ mg/L} \\ &= 0.5 \text{ mg/L} \end{aligned}$$

b. Determine available fluoride ion (F^-):

From Table B-1, pure sodium fluorosilicate has 60.7 percent available fluoride ion.

c. Calculate dry chemical feed rate using Eq B-2:

$$\text{Dry chemical feed rate (lb/day)} = \frac{2.88 \text{ mgd} \times 0.5 \text{ mg/L} \times 8.34 \text{ lb/gal}}{0.607 \times 0.98 \times 1{,}440 \text{ min/day}}$$
$$= 0.014 \text{ lb/min}$$

d. Calculate chemical solution feed rate using Eq B-3:

$$\text{Chemical solution feed rate (gal/hr)} = \frac{0.014 \text{ lb/min}}{8.34 \text{ lb/gal} \times 0.002} \times 60 \text{ min/hr}$$
$$= 50.4 \text{ gal/hr}$$

Example 2. Sodium fluorosilicate with a commercial purity of 95 percent is to be fed. If the plant pumping rate is 4 MLD, determine the dry chemical feed rate in kilograms per day and the chemical solution feed rate in milliliters per minute. The natural fluoride present in the water is insignificant.

Given:

Chemical purity = 95 percent
Pumping rate = 4 MLD
Chemical solution density = 999.4 kg/m^3
Chemical solution strength = 0.2%
Natural fluoride concentration = 0 mg/L

Solution:

a. Determine optimal fluoride concentration:
From Table B-1, the optimal fluoride concentration is 0.9 mg/L.

b. Determine fluoride dose:

$$\text{Dose (mg/L)} = \text{optimal fluoride level (mg/L)} - \text{natural fluoride level (mg/L)}$$
$$= 0.7 \text{ mg/L} - 0.0 \text{ mg/L}$$
$$= 0.7 \text{ mg/L}$$

c. Determine available F^-:
From Table B-1, pure sodium fluorosilicate has 60.7 percent available fluoride ion.

d. Calculate dry chemical feed rate using Eq B-2:

$$\text{Dry chemical feed rate (kg/min)} = \frac{4 \text{ MLD} \times 0.70 \text{ mg/L}}{0.607 \times 0.95 \times 1{,}440 \text{ min/day}}$$
$$= 0.0034 \text{ kg/min}$$

e. Calculate chemical solution feed rate using Eq B-3:

$$\text{Chemical feed rate (L/hr)} = \frac{0.0034 \text{ kg/min}}{999.4 \text{ kg/m}^3 \times 0.002} \times 1{,}000 \text{ L/m}^3 \times 60 \text{ min/hr}$$
$$= 102.1 \text{ L/hr}$$

Hydrofluorosilicic Acid

Example 1. The optimal fluoride concentration has been determined to be 0.7 mg/L. The solution strength of commercial hydrofluorosilicic acid to be fed is 23 percent, and the solution density is 10 lb/gal. The pumping rate is 4,000 gpm, and the natural fluoride present is insignificant. Determine the feed rate in gallons per hour.

Given:

Optimal fluoride concentration = 0.7 mg/L
Natural fluoride concentration = 0.0 mg/L
Pumping rate = 4,000 gpm = 5.76 mgd
Chemical solution density = 10 lb/gal
Chemical solution strength = 23 percent

Solution:

a. Determine fluoride dosage:

$$\text{Dosage (mg/L)} = \text{optimal fluoride level (mg/L)} - \text{natural fluoride level (mg/L)}$$
$$= 0.7 \text{ mg/L} - 0.0 \text{ mg/L}$$
$$= 0.7 \text{ mg/L}$$

b. Determine available F^-:

From Table B-2, pure hydrofluorosilicic acid has 79.2 percent available fluoride ion.

c. Calculate feed rate using Eq B-4:

$$\text{Chemical solution feed rate (gal/hr)} = \frac{5.76 \text{ mgd} \times 0.7 \text{ mg/L} \times 8.34 \text{ lb/gal}}{10 \text{ lb/gal} \times 0.792 \times 0.23 \times 24 \text{ hr/day}}$$
$$= 0.8 \text{ gal/hr}$$

Example 2. The optimal fluoride concentration has been determined to be 0.7 mg/L. The solution strength of commercial hydrofluorosilicic acid to be fed is 23 percent, and the solution density is 1,198.3 kg/m³. The pumping rate is 8 MLD, and the natural fluoride present is insignificant. Determine the feed rate in liters per hour.

Given:

Optimal fluoride concentration =	0.7 mg/L
Natural fluoride concentration =	0.0 mg/L
Pumping rate =	8 MLD
Chemical solution density =	1,198.3 kg/m³
Chemical solution strength =	23 percent

Solution:

a. Determine fluoride dose:

$$\begin{aligned}\text{Dose (mg/L)} &= \text{optimal fluoride level (mg/L)} - \text{natural fluoride level (mg/L)}\\ &= 0.7\text{ mg/L} - 0.0\text{ mg/L}\\ &= 0.7\text{ mg/L}\end{aligned}$$

b. Determine available F^-:

From Table B-2, pure hydrofluorosilicic acid has 79.2 percent available fluoride ion.

c. Calculate feed rate using Eq B-4:

$$\begin{aligned}\text{Chemical solution feed rate (L/hr)} &= \frac{8\text{ MLD} \times 0.7\text{ mg/L}}{1{,}198.3\text{ kg/m}^3 \times 0.792 \times 0.23 \times 24\text{ hr/day}} \times 1{,}000\text{ L/m}^3\\ &= 1.07\text{ L/hr}\end{aligned}$$

Sodium Fluoride

Example 1. The fluoride dosage has been determined to be 0.7 mg/L. Sodium fluoride (commercial purity of 95 percent) is to be added to a saturator and the saturated solution used to treat a plant rate of 2 mgd. Daily plant production is 1.5 million gallons. Determine the daily chemical usage in pounds per day and the chemical feed rate in gallons per hour.

Given:

Fluoride dosage =	0.7 mg/L
Plant rate =	2 mgd
Daily production =	1.5 million gallons
Commercial purity =	95 percent

Solution:

a. Determine available F^-:

From Table B-2, pure sodium fluoride has 45.25 percent available fluoride ion.

b. Calculate daily chemical usage using Eq B-1:

$$\text{Chemical usage (lb/day)} = \frac{0.7 \text{ mg/L} \times 1.5 \text{ million gallons} \times 8.34 \text{ lb/gal}}{0.4525 \times 0.95}$$
$$= 20 \text{ lb/day}$$

c. Calculate chemical feed rate using Eq B-7:

$$\text{Chemical feed rate} = \frac{2 \text{ mgd} \times 0.7 \text{ mg/L}}{18{,}000 \text{ mg/L}}$$
$$= 7.77 \times 10^{-5} \text{ mgd}$$

d. Convert chemical feed rate to gallons per hour:

$$\text{Chemical feed rate} = \frac{7.77 \times 10^{-5} \text{ mgd}}{24 \text{ hr/day}} \times 1{,}000{,}000 \text{ gal/million gal}$$
$$= 3.2 \text{ gal/hr}$$

Example 2. The fluoride dosage has been determined to be 0.7 mg/L. Sodium fluoride (commercial purity of 95 percent) is to be added to a saturator and the saturated solution used to treat a plant rate of 10 MLD. Daily plant production is 6 megaliters. Determine the daily chemical usage in kilograms per day and the chemical feed rate in liters per hour.

Given:

Fluoride dosage = 0.7 mg/L
Plant rate = 10 MLD
Daily production = 6 million liters
Commercial purity = 95 percent

Solution:

a. Determine available F^-:

From Table B-2, pure sodium fluoride has 45.25 percent available fluoride ion.

b. Calculate daily chemical usage using Eq B-1:

$$\text{Chemical usage (kg/day)} = \frac{0.7 \text{ mg/L} \times 6 \text{ million liters}}{0.4525 \times 0.95}$$
$$= 9.8 \text{ kg/day}$$

c. Calculate chemical feed rate using Eq B-7:

$$\text{Chemical feed rate} = \frac{10\text{ MLD} \times 0.7\text{ mg/L}}{18{,}000\text{ mg/L}}$$
$$= 3.9 \times 10^{-4}\text{ MLD}$$

d. Convert chemical feed rate to liters per hour:

$$\text{Chemical feed rate} = \frac{3.9 \times 10^{-4}\text{ MLD}}{24\text{ hr/day}} \times 1{,}000{,}000\text{ L/ML}$$
$$= 16.3\text{ L/hr}$$

EXAMPLE CALCULATED DOSAGE PROBLEMS

Sodium Fluorosilicate

Example 1. A plant uses 50 pounds of sodium fluorosilicate in treating 5,540,000 gallons of water (assume 98.5 percent chemical purity).

The solution for determining the calculated dosage (using Eq B-5):

$$\text{Calculated concentration (mg/L)} = \frac{50\text{ lb} \times 0.607 \times 0.985}{5.54\text{ million gallons} \times 8.34\text{ lb/gal}}$$
$$= 0.65\text{ mg/L}$$

Example 2. Calculate the dosage for a plant that uses 6 kg sodium fluorosilicate in treating 6,800,000 liters of water (assume 98.5 percent chemical purity).

$$\text{Calculated concentration (mg/L)} = \frac{6\text{ kg} \times 0.607 \times 0.985}{6.8\text{ million liters}}$$
$$= 0.53\text{ mg/L}$$

Hydrofluorosilicic Acid

Example 1. What is the calculated dosage for a plant using 3.5 gallons of 23 percent hydrofluorosilicic acid with a solution density of 10 lb/gal to treat 1,226,000 gallons of water?

Solution (using Eq B-6):

$$\text{Calculated concentration (mg/L)} = \frac{3.5\text{ gal} \times 10\text{ lb/gal} \times 0.23 \times 0.792}{1.226\text{ million gallons} \times 8.34\text{ lb/gal}}$$
$$= 0.62\text{ mg/L}$$

Example 2. Determine the dosage for a plant that uses 20 liters of 23 percent hydrofluorosilicic acid with a solution density of 1,198.3 kg/m³ in treating 6,500,000 liters of water.

Solution (using Eq B-6):

$$\text{Calculated concentration (mg/L)} = \frac{20\ \text{L} \times 1{,}198.3\ \text{kg/m}^3 \times 0.23 \times 0.792}{6.5\ \text{ML} \times 1{,}000\ \text{L/m}^3}$$

$$= 0.67\ \text{mg/L}$$

Sodium Fluoride

Example 1. What is the calculated dosage for a plant using 7 gal from its saturator to treat 200,000 gal of water?

Solution (using Eq B-8):

$$\text{Calculated concentration (mg/L)} = \frac{7\ \text{gal} \times 18{,}000\ \text{mg/L}}{200{,}000\ \text{gal}}$$

$$= 0.63\ \text{mg/L}$$

Example 2. Determine the calculated dosage for a plant using 50 liters of solution from its saturator to treat 1,360,000 liters of water.

Solution (using Eq B-8):

$$\text{Calculated dosage (mg/L)} = \frac{50\ \text{L} \times 18{,}000\ \text{mg/L}}{1{,}360{,}000\ \text{L}}$$

$$= 0.66\ \text{mg/L}$$

SIMPLIFICATION OF FEED RATE AND DOSAGE CALCULATIONS

Calculations can be simplified by using charts or tables. The figures in the column at the extreme right of Table B-3 can be used for calculating the amount of fluoride compound to add for quantities of water other than 1 million gallons. For example, the amount of 97 percent sodium fluoride needed for 10,000 gallons would be 10,000/1,000,000, or 0.01, times the amount indicated. For 2 million gallons, the amount of fluoride compound needed would be twice as much, and so forth.

Similarly, if a concentration of F^- other than 0.7 mg/L is to be added (because of the presence of natural fluoride or an optimal concentration other than 0.7 mg/L), multiplying the figures in the right-hand column by the appropriate factor will give the number of pounds to use. For example, if 0.3 mg/L F^- occurs naturally and the optimal concentration is 0.7 mg/L, only 0.4 mg/L would have to be added, or one half as much as indicated.

Table B-2 Amount of additive needed based on purity

Compound	Available F^- (lb F/lb compound)	Commercial Purity (%)	Compound Required per million gallons of water for 1 mg/L (*lb*)	Compound Required per million liters of water for 1 mg/L (*kg*)
Sodium fluoride	0.4525	95	19.4	2.33
		96	19.2	2.30
		97	19.0	2.28
		98	18.8	2.26
		99	18.6	2.23
Sodium fluorosilicate	0.607	95	14.5	1.73
		96	14.3	1.72
		97	14.2	1.70
		98	14.0	1.68
		98.5	13.9	1.67
		99	13.9	1.66
Hydroflurosilicic acid	0.792	20	52.7	6.31
		21	50.1	6.01
		22	47.9	5.74
		23	45.8	5.49
		24	43.9	5.26
		25	42.1	5.05
		26	40.5	4.86
		27	39.0	4.68
		28	37.6	4.51
		29	36.3	4.35
		30	35.1	4.21

Appendix C

Frequently Asked Questions

Is fluoride harmful to human health?

The weight of evidence from all available peer-reviewed scientific research does not support a link between exposure to fluoride in drinking water at the proposed recommended level of 0.7 mg/L and any significant adverse health effects.

Does fluoride only help children?

While fluoride is incorporated into tooth enamel during childhood, fluoride continues to provide protection against dental decay throughout a person's life by helping to enhance incorporation of calcium and phosphate back into the tooth enamel (remineralization).

Does the USEPA mandate the addition of fluoride for dental health?

Fluoridation of public water supplies is not mandated by the US Environmental Protection Agency (USEPA) or any other federal agency in the United States. The 1974 Safe Drinking Water Act (SDWA) specified that no national primary drinking water regulation can require the addition of any substance for preventive health benefits not related to drinking water contamination. This prohibition inherently established fluoridation as a decision to be made by each individual state or local municipality.

The SDWA further required that the USEPA determine the level of contaminants in drinking water at which no adverse health effects are likely to occur, and to apply limits based on possible health risks assuming a lifetime of exposure. Fluoride was regulated in 1992 under Phase II of the Chemical Contaminants Rules. The maximum contaminant level goal (MCLG) and maximum contaminant level (MCL) for fluoride are both 4 mg/L (4 ppm). This is the level below which a lifetime of exposure is not expected to cause health problems, and it is an enforceable maximum allowable concentration under the SDWA. The secondary MCL for fluoride is 2 mg/L and is meant to prevent discoloration or pitting of teeth in children exposed to water with high naturally occurring levels of fluoride during their formative years. A state may choose to adopt a secondary standard (or any level lower than the federal MCL) as its enforceable standard.

If the USEPA does not mandate water fluoridation, where do fluoridation directives come from and how is fluoridation regulated?

Although only 13 states have a fluoridation mandate, no state prohibits fluoridation. In most states, the decision is made at a local (regional or municipal) level. The decision at a local level can arise and be implemented in several ways. As an example, a local municipal health agency may make the recommendation to a city's governing body (i.e., mayor, city council, etc.). The local legislature may put the recommendation to referendum as a ballot question during a future election. Some municipalities have local ordinances requiring fluoridation. A local ordinance may be instituted if the legislature grants the recommending health agency the authority to require fluoridation without a referendum. The issue can become more complex when a public water supplier serves more than one municipality, such as in consecutive systems.

In most cases, the water supplier is not the entity that authorizes fluoridation, although it is tasked with implementation. The mandate to fluoridate can be the result of a legislative act or a directive from a state or local health agency. Regardless of the mechanism by which fluoridation was instituted, the water supplier is responsible for working with its primary regulatory drinking water administrator to ensure compliance with relevant permit requirements, such as design and maintenance of the fluoride chemical feed system, monitoring fluoride levels in water samples, reporting data, and public notification.

The obligation a utility may have under fluoridation requirements is likely to be different in different communities. It is imperative that the water utility in any community considering fluoridation consult with its state or regional drinking water administrator throughout the process of consideration and implementation of a fluoridation program.

When and where was water fluoridation first practiced in the United States?

On Jan. 25, 1945, Grand Rapids, Mich., became the first city to be artificially fluoridated. In this study, which was projected to last 15 years, the oral health in Grand Rapids was compared with that of Muskegon, a nearby community consuming water from the same source, and Aurora, a similar community with naturally occurring optimum fluoride levels. Six years after the study began, surveys indicated that decay levels in six-year-old children (i.e., those born since fluoridation commenced) in Grand Rapids was almost half that of Muskegon. In July 1951, city officials in Muskegon decided to withdraw from the study and fluoridate the town's water supply. Similar results were observed in the other trials referenced above, i.e., significant reduction in dental decay rates were described in the fluoridating towns, with little or no change in the controls. Since that time, the US population receiving optimally fluoridated drinking water has continued to increase.

How extensive is the practice of water fluoridation in the United States currently?

As of 2010, approximately 204.3 million people in the United States (74 percent of the population served by public water systems) received optimally fluoridated water. This is approximately 66 percent of the total US population. Another 10 million Americans are served by community water systems with naturally occurring fluoride.

What is the best location to add fluoride in the treatment process?

In a water treatment plant, fluoride is best added as late in the treatment process as possible to avoid chemical losses through processes such as coagulation and filtration. In all cases, it must be added to the entire flow in a common pipe or channel.

What form of fluoride should be used?

A utility should determine its chemical needs (i.e., quantity), evaluate what chemical form it is best able to handle (e.g., liquid or solid), evaluate what supplies are available locally, and estimate what chemical costs are projected to be in order to assist with this decision.

At what concentration point should the feed be discontinued?

The Centers for Disease Control and Prevention (CDC) recommends the fluoride feed be discontinued when the concentration exceeds 4.0 mg/L until the issue causing the overfeed condition is found and corrected.

How often and where should samples be collected to monitor for fluoride and ensure that the fluoride chemical feed system is operating properly and in control?

For a water system that adds fluoride in treatment, there may be monitoring requirements in addition to the standard SDWA requirements. These requirements are a stipulation within operating permits as related to the addition of fluoride. Because this monitoring requirement is specifically related to the addition of fluoride, the monitoring requirement will be more frequent (i.e., daily) than most SDWA compliance monitoring requirements. The sampling locations may be the same as required for SDWA compliance monitoring, or they may include different or additional locations, including locations in treatment prior to the addition of fluoride. These requirements vary from place to place, so it is imperative that fluoridating water systems ensure they are in compliance with operational monitoring requirements as imposed by their regulatory agency.

Where can I find more information about public-notification requirements in the event of an accidental fluoride overfeed that causes fluoride levels in treated water to exceed the MCL?

Subpart D in the Code of Federal Regulations (CFR) at part 141.32 of the SDWA requirements provides information about public notification requirements when MCLs are exceeded. Individual states may have requirements that are more stringent than the federal requirements, so water systems need to be aware of what is required by their regulatory agency.

This page intentionally blank.

Appendix D

Selected Sources of Scientific Water Fluoridation Information

The following is a selected list of additional sources of scientific information on water fluoridation that have been peer reviewed or are products of the public health organizations of the several countries from which they originate. Be aware that when searching the Internet for water fluoridation information, the most visited sites are listed first, and they typically belong to organizations opposed to water fluoridation. All major scientific and public health organizations in the United States and in many other countries are in strong support of water fluoridation. The sites listed here present the scientific data and positions of these organizations.

The following links are from the US Surgeon General, the US Centers for Disease Control and Prevention, the US National Research Council, and the American Dental Association.

- American Dental Association *Fluoridation Facts* presents water fluoridation information by subject, and includes citations from 358 peer-reviewed articles on water fluoridation. This is the best single source of information for those new to the subject of water fluoridation: Fluoridation Facts Compendium
- US Surgeon General: Statement on Community Water Fluoridation (2004)
- US Centers for Disease Control and Prevention: Water Fluoridation Information Index Page
- US Surgeon General's *Report on Oral Health in America*, the first comprehensive report on Oral Health in America. Information on water fluoridation is found on pages 158–162: http://profiles.nlm.nih.gov/ps/retrieve/ResourceMetadata/NNBBJT/

- US Centers for Disease Control and Prevention: Ten Great Public Health Achievements – United States, 1900–1999
- US National Research Council, *Fluoride in Drinking Water,* analyzes recent peer-reviewed scientific research on various health effects from exposure to fluoride: Fluoride in Drinking Water

The following links are from US states, Canadian provinces, and other public health websites.

- California Department of Public Health: Benefits of Fluoridation and Fluoride Facts & Fiction
- Canadian Dental Association, Frequently Asked Questions on Fluoridation: FAQ
- Community Guide, Preventing Dental Caries: Community Water Fluoridation
- Alberta Health Services, Fluoridation Article, Water Fluoridation — What Does the Science Say?
- Water Fluoridation Article, QuackWatch.com: Don't Let the Poisonmongers Scare You!
- Delta Dental Plans Association Article on Bottled Water, Fluoride and Cavities: Bottled Water: Cause of Cavity Comeback?

The following links are from public health agencies in countries outside of North America that are practicing water fluoridation.

- Australian Dental Association (Queensland Branch): Position Statement
- British Fluoridation Society: Main Page
- New Zealand Public Health Association: Policy Statement

The following links are from the Pan American Health Organization and the Salt Fluoridation Program in Mexico, and describe the practice of salt fluoridation.

- Pan American Health Organization (PAHO) Salt Fluoridation Activities, described in an article at Dental Plans.Com, Inc.: Salt Fluoridation
- National Coordinator for the Salt Fluoridation Program in Mexico: Article on Salt Fluoridation; reported by the Network for Cooperation in Studies and Development of Dental Resources to the Health Sector (CEDROS; Federal University of Rio de Janeiro): Salt Fluoridation in Mexico

The following is a list of public health organizations in strong support of water fluoridation.

- American Academy of Pediatrics
- American Academy of Pediatric Dentistry
- American Association for the Advancement of Science
- American Association for Dental Research
- American Association of Dental Schools
- American Association of Public Health Dentistry
- American College of Dentists
- American Council on Science and Health

- American Dental Assistants Association
- American Dental Association
- American Dental Hygienists Association
- American Dietetic Association
- American Federation of Labor and Congress of Industrial Organizations
- American Hospital Association
- American Institute of Nutrition
- American Medical Association
- American Osteopathic Association
- American Pharmaceutical Association
- American Public Health Association
- American Public Welfare Association
- American School Health Association
- American Society of Clinical Nutrition
- American Society for Dentistry for Children
- American Society for Nutritional Sciences
- American Veterinary Medical Association
- American Water Works Association
- Association for Academic Health Centers
- Association of State and Territorial Dental Directors
- British Dental Association
- British Fluoridation Society
- British Medical Association
- California Children's Services
- California Dental Association (jointly with US Public Health Service)
- California Dental Hygienists' Association
- California Department of Health Services
- California Fluoridation Task Force
- California Medical Assistance Commission
- California State Assembly
- California State PTA
- Center for Science in the Public Interest
- Centers for Disease Control and Prevention
- Centro de Niños
- Children Now
- Children's Dental Foundation
- Children's Hospital Los Angeles
- Children's Roundtable
- Chinese American Dental Society
- Clinica Para Las Americas
- Community Health Councils Project
- Department of National Health and Welfare (Canada)

- Delta Dental Plans Association
- Dental Coalition for Needy Children
- Dental Health Foundation
- European Organization for Caries Research
- Federation of Community Coordinating Councils of LA County
- Federation Dentaire Internationale
- Food and Nutrition Board
- Foundation for Children's Dental Health
- Great Britain Ministry of Health
- Harbor Dental Society
- Health and Safety/Health Care Workers
- Health Insurance Association of America
- Health League of Canada
- Healthy LA 2000 Council
- Hispanic Dental Association
- Indian Dental Association
- Indian Health Service
- Inter-Association Committee on Health
- International Association for Dental Research
- Japanese-American Dental Society
- Kaiser Permanente
- Korean Dental Society
- LA 4 Kids
- Los Angeles Citizens for Better Dental Health
- Los Angeles City/County Indian Commission
- Los Angeles County Children's Planning Council
- Los Angeles County Medical Association
- Los Angeles Free Clinic
- Latin American Dental Association
- League of California Cities
- Long Beach Water Department
- Los Angeles Association of Women Dentists
- Los Angeles Chamber of Commerce
- Los Angeles County Board of Supervisors
- Los Angeles County Department of Health Services
- Los Angeles Dental Society
- Los Angeles Department of Children and Family Services
- Los Angeles Grand Jury Report
- Los Angeles Mayor's Committee on Children, Youth and Families
- Los Angeles Oral Health Foundation
- Los Angeles PTA
- Los Angeles Roundtable for Children

- Los Angeles Unified School District
- Mayo Clinic
- Mexican American Opportunity Foundation Head Start
- National Academy of Sciences
- National Alliance for Oral Health
- National Cancer Institute
- National Confectioners Association
- National Congress of Parents and Teachers
- National Health Council
- National Institute of Dental Research
- National Institutes of Health
- New York Academy of Medicine
- Office of Dental Health Services
- Older Women's League
- Oral Health in Inglewood
- Pan American Health Organization
- Policy & Development Association
- Project Heavy – West (Los Angeles)
- Public Council
- Public Health Commission
- Public Health Programs and Services
- Queen of Angels – Hollywood Presbyterian Hospitals
- Royal College of Physicians (London)
- San Fernando Valley Dental Society
- San Gabriel Unified School District
- San Gabriel Valley Dental Society
- South Bay Children's Health Center, Inc.
- South Bay Free Clinic
- Southern California Filipino Dental Society
- Southern California Public Health Association
- The Children's Dental Center
- The Dental Health Foundation (of California)
- Travelers Insurance Company
- UC San Francisco, School of Dentistry
- UCLA Community Partnership
- UCLA REI WIC Program
- UCLA School of Public Health
- University of San Diego Medical Center
- US California School of Dentistry
- US Department of Defense
- US Department of Health and Human Services
- US Department of Veterans Affairs

- US Junior Chamber of Commerce
- US Public Health Service
- Western Center on Law and Poverty
- Western Los Angeles Dental Society
- World Health Organization
- Youth Law Center

Appendix E

Inspection Checklists

This section provides inspection checklists for the three types of fluoride systems. Items in each list are presented in alphabetical order. While these checklists are not intended to be exhaustive, they nevertheless comprise the majority of items typically included in the three types of fluoridation systems.

1. Sodium Fluoride Saturator Metering System

- Alarms (visual and/or audio)
  - Containment area
  - High/low fluoride concentration in finished water
- Auxiliary equipment
  - Unions on pipelines
  - Vacuum breaker to saturator
- Backflow prevention devices on makeup water line
  - Reduced pressure backflow preventer (type, model, height off floor, location, and size approved by state or local regulations)
- Containment volume at 110% of saturator tank
- Controllers
- Cross connections protected
- Dust collectors and wet scrubbers
- Eductors
- Floor drains (if allowed by state)
  - Ultimate disposal site
- Fluoride injection point(s)
  - Location
  - Number
  - Type (corporation stop/injection nozzle or basin diffuser pipe)

- Flow pacing of metering pump(s)
  - (4–20 mA electrical signal)
  - Pacing meter
  - Current to frequency converter
- Inline water mixer
  - Location
  - Type
- Lab testing equipment
  - Location
  - SPADNS fluoride reagent
  - Electrode
  - Continuous analyzer
- Location of saturator and metering pump
  - Secure from tampering and vandalism
- Makeup water to saturator
  - Hardness to be removed if over 50–75 mg/L as $CaCO_3$ (from AWWA Standard B701-11)
    - Hardness reduction filter
    - Zeolite softener
  - Pressure reducing valve with gauge installed if over 80 psi (from AWWA Standard B701-11)
  - Quality
- Water meter with totalizer on saturator makeup water line to saturator
  - Location
  - Size
  - Strainer included
  - Units of measurement (gallons, cubic feet, liters)
- Metering pump interlock using a flow switch
  - Location (recommend upstream of fluoride injection point)
  - Type
- Metering pumps and feeders
  - Type
  - Number
  - Spares
  - Location
  - Clear PVC calibration chamber (minimum 100 mL) with suction pipe/tubing for metering pump
  - Plug and receptacle for metering pump of 115 VAC twist type (help prevent overfeeds)
  - Anti-siphon valves (minimum of two on discharge line and accessibly located for observation and testing)

- Operator safety equipment*
  - Personal list of four safety items (from AWWA Standard B701-11)
  - Exhaust fans
  - Room location
  - Eyewash
- Piping and tubing
  - Type
  - Pressure rating
  - Hangers
  - ID labeled
- Refractometer (to field test percent NaF saturation solution from top of saturator tank)
- Sample tap (before first user)
  - Type
  - Inline mixer
  - Location (inside or outside building)
  - ID label installed
  - Yard hydrant
- Saturator
  - Type (upflow or downflow), upflow type preferred
  - Size in gallons (50, 70, 110, or other size smaller)
  - Location (secure from tampering and vandalism)
  - Construction materials
  - Tank cover lid
  - Vacuum breaker (non-pressure type) on makeup water supply
  - Built-in momentary light to observe NaF chemical level
  - Makeup water flow restrictor (2 gpm maximum)
  - Level fill system
  - Containment and floor drains
  - Saturator overflow prevention system (as solenoid valve or liquid level switch may fail in open position)
  - Markings on outside of saturator tank to indicate chemical depth, full and refill levels
- Signage
  - Standard operating procedure (SOP)
  - Saturator tank label (if state right-to-know law prevails)

---

* The following protective clothing and equip¬ment should be the minimum available:

1. A National Institute for Occupational Safety and Health/Mine Safety Health Administration (NIOSH/MSHA) approved, high-efficiency dust respirator (chemical mask) with a soft rubber face-to-mask seal and replaceable cartridges.
2. Gauntlet neoprene gloves (12-in. [300-mm] minimum glove length).
3. Heavy-duty neoprene aprons.
4. Splash-proof goggles.

2. Sodium Fluorosilicate Metering System

- Alarms (visual and/or audio)
  - Containment area
  - High/low fluoride concentration in finished water
- Auxiliary equipment
  - Unions
  - Strainers
- Backflow prevention devices on makeup water line
  - Reduced pressure backflow preventer (type, model, height off floor, location, and size approved by state or local regulations)
- Bag loaders
- Containment
- Controllers
- Cross connections protected
- Dissolving tank float control system
- Dissolving tanks
  - Size
  - Overflow
  - Material
- Dry feeder
  - Gravimetric
  - Volumetric
  - Control panel
  - Capacity in cubic feet or cubic meters
- Dry feeder/metering pump interlock/flow switch
- Dust collectors and wet scrubbers
- Eductors
- Exhaust fans
  - Number of room changes per hour
  - Location
  - Operation
  - Switch location
- Floor drains (if allowed by state)
  - Ultimate disposal
- Flow pacing of dry feeder
  - (4–20 mA electrical signal)
  - Pacing meter
  - Current to frequency converter
- Fluoride injection point(s)
  - Location
  - Number
  - Type (corporation stop/injection nozzle or basin diffuser pipe)

- Hopper agitators
- Hoppers
- Inline water mixer
  - Location
  - Type
- Lab testing equipment
  - Location
  - SPADNS fluoride reagent
  - Electrode
  - Continuous analyzer
- Location of dry feeder
  - Secure from tampering and vandalism
- Makeup water to dry feeder
  - Softener
  - Rotameter
  - Temperature
  - Water quality
  - Pressure
- Metering pump (if needed in case gravity feed to injection point is not practical)
  - Clear PVC calibration chamber (minimum 100 mL) with suction pipe/tubing for metering pump
  - Number
  - Location
  - Anti-siphon valves (minimum of two on discharge line and accessibly located for observation and testing)
- Mixers for dissolving tank
  - Number
  - Type
  - Operation
  - Construction materials
  - RPM
- Operator safety equipment
  - Personal list of four safety items (from AWWA Standard B702-11)*
  - Room location
  - Eyewash (and safety shower if required by authority having jurisdiction)

---

* The following list of protective clothing and equipment should be the minimum available:
  1. A high-efficiency dust respirator (chemical mask) with a soft rubber face-to-mask seal and replaceable cartridges approved by the National Institute for Occupational Safety and Health/Mine Safety Health Administration (NIOSH/MSHA).
  2. Gauntlet neoprene gloves (12-in. [300-mm] minimum glove length).
  3. Heavy-duty neoprene aprons.
  4. Splash-proof goggles.

- Piping and tubing
  - Type
  - Hangers
- Sample tap (before first user)
  - Type
  - Inline mixer
  - Location (inside or outside building)
  - ID label installed
  - Yard hydrant
- Scales
  - Type
  - Capacity
  - Accuracy
- Signage
  - SOP
  - Dry feeder
  - Labels
- Valves and meters
- Weight chart recorders on scale

3. Fluorosilicic Acid Metering System

- Alarms (visual and/or audio)
  - Containment area
  - High/low fluoride concentration in finished water
- Auxiliary equipment
  - Unions
  - Strainers
- Backflow prevention devices on makeup water line (if used to dilute pumped fluorosilicic acid)
  - Reduced pressure backflow preventer (type, model, height off floor, location, and size approved by state or local regulations)
- Backflow prevention devices
- Bulk tank and day tank vents to outside
  - Diameter
  - Rise continuously upward
  - Outside screen type and material
  - Material
  - Hangers
  - Outside location (away from windows, parked vehicles, walkways)
- Bulk tank
  - Number
  - Location

- Foundation
- Type
- Manway
- Construction (single/double wall)
- Fill pipe, vent, and overflow test air/water
- Fill station

- Bulk tank overflow pipe
  - Diameter
  - Method to prevent from acting as inside vent
- Bulk tank level measuring system
  - Sight tube
  - Ultrasonic
- Carry water for acid dilution after pumping (optional)
  - Protected with appropriate backflow device
  - Pressure
  - Water meter
  - Check valves
  - Interlock
- Containment volume at 110% of bulk tank(s) and day tank(s)
- Controllers
- Cross connections protected
- Day tank
  - Construction (single/double wall)
  - Fill pipe
  - Overfill prevention system
  - Vent (separate from bulk tank vent)
  - Light to observe chemical depth
  - Flexible joints in horizontal plane for fill, vent, and suction line (use tubing)
- Eductors
- Exhaust fans
  - Number of room changes per hour
  - Location
  - Operation
  - Switch location
- Floor drains (if allowed by state)
  - Ultimate disposal
- Flow pacing of metering pump(s)
  - (4–20 mA electrical signal)
  - Pacing meter
  - Current to frequency converter
- Fluoride injection point(s)
  - Location

  - Number
  - Type (corporation stop/injection nozzle or basin diffuser pipe)
- Hauling equipment for drums, barrels, or carboys
  - Type
  - Containment restrictions
- Hydrometer (to field test percent fluosilicic acid)
- Inline water mixer
  - Location
  - Type
- Lab testing equipment
  - SPADNS fluoride reagent
  - Electrode
  - Location
  - Continuous analyzer
- Location of tank(s) and pumps
  - Secure from tampering and vandalism
- Metering pump interlock/flow switch
- Metering pumps and feeders
  - Operation
  - Number
  - Clear PVC calibration chamber (minimum 100 mL) with suction pipe/tubing for metering pump
  - Plug and receptacle for metering pump of 115 VAC twist type (help prevent overfeeds)
  - Anti-siphon valves (minimum of two on discharge line and accessibly located for observation and testing)
- Operator safety equipment
  - Personal list of four safety items (from AWWA Standard B703-11)*
  - Room location
  - Eyewash and shower
  - 2.5 percent calcium gluconate gel in a tube for areas suspected of direct fluorosilicic acid skin contact
- Piping and tubing
  - Type
  - Hangers
- Sample tap (before first user)
  - Type

---

* The following list of protective clothing and equipment should be the minimum available:

1. Gauntlet neoprene gloves (12-in. [300-mm] minimum glove length).
2. Full 8-in. (200-mm) face shield and chemical splash-proof safety goggles.
3. Heavy-duty, acid-proof-type neoprene aprons.
4. Safety shower and eyewash in an easily accessible location.

- Inline mixer
  - Location (inside or outside building)
  - ID label installed
- Scales
  - Type
  - Capacity
  - Accuracy
- Signage
  - SOP
  - Labels
  - Fill station for bulk tanks
- Transfer pump
  - Number
  - Type
  - Location
  - Control system
  - Siphon protection method
- Valves and meters
- Weight chart recorders on scale
- Yard hydrant

Appendix F

Testing Checklists

This section provides testing checklists for the three types of fluoride systems. Each checklist includes most items typically tested in the corresponding type of fluoridation system, although the checklists are not intended to be exhaustive.

1. Sodium Fluoride Metering System

1. Test completed fluoride system with water to test for leaks and to set approximate fluoride pump feed rate (percent strokes and strokes per minute). Run disinfected water through the saturator for one to two days and record makeup water readings and total gallons of water used. This step does not require any sodium fluoride to be added to saturator yet.
2. Test fluoride lab tester after calibration with fluoride standards.
3. Run fluoridated water to waste (such as through a valved-off hydrant) until testing is complete.
4. Test fluoridated metering pump interlock/flow switch system.
5. Test fluoride continuous analyzer by adding a known fluoridated sample to test pipe inlet if one is installed.
6. Verify that sodium fluoride is used.
7. Test all alarm systems (high and low).
8. Test all piped safety showers and eyewashes.
9. Prepare a metering pump calibration curve for optimum fluoride levels (water flow rate vs. metering pump settings).
10. Test saturator overflow prevention system (observe only during saturator fill process).
11. Test SCADA system to verify a decrease and increase in pacing signal works properly.
12. Keep accurate records of all testing results.

2. Sodium Fluorosilicate Metering System

1. Test all alarm systems (high and low).
2. Test fluoride lab tester.
3. Test room air exhaust system.
4. Test all piped safety showers and eyewashes.
5. Run fluoridated water to waste (such as through a valved-off hydrant) until testing is complete.
6. Test fluoride continuous analyzer by adding a known fluoridated sample to test pipe inlet.
7. Add clean water to dissolving tank using a hose and perform pressure test on tank. (If the tank is gravity fed, this test cannot be done.) It may be necessary to first disinfect the water used in testing.
8. Verify that sodium fluorosilicate is used.
9. Test scale accuracy with known weights.
10. Test hopper dust collection system.
11. Test fluoridated dry feeder/metering pump interlock/flow switch system.
12. Prepare a dry feeder/metering pump calibration curve for optimum fluoride levels (water flow rate vs. dry feeder/metering pump settings).
13. Test SCADA system to verify a decrease and increase in pacing signal works properly.
14. Keep accurate records of all testing results.

3. Fluorosilicic Acid Metering System

1. Fill up bulk tank with clean water and test for leaks. One option is to connect a hose using fill pipe connection so 2-in. fill pipe system can be checked for leaks as well as tank overflow pipe.
2. Consult tank manufacturer for proper method of testing tank vent system and related pipe joints. One option may be to introduce very low pressure air supplied by portable compressor with calibrated 0–15 psi or 0–30 psi pressure gauge. Any joint leaks may be observed from presence of air bubbles appearing in a soap detection solution using a recommended two-person test operation crew. Follow all safety precautions for air testing as this test method can create unique hazards from dynamic piping failures.
3. Fill up day tank system, if applicable, with clean water and test for leaks.
4. Test transfer pump system, if applicable, using clean water from bulk tank to fill day tank.
5. Test fluoridated metering pump interlock/flow switch system.
6. Test all alarm systems (high and low).
7. Test fluoride lab tester.
8. Test all piped safety showers and eyewashes.
9. Test room air exhaust system.

10. Test fluoride system with water (run disinfected water through the bulk tank and day tank for one to two days and observe water scale readings).
11. Run fluoridated water to waste (such as through a valved-off hydrant) until testing is complete.
12. Test fluoride continuous analyzer by adding a known fluoridated sample to test pipe inlet if one is installed.
13. Verify that 23 percent to 30 percent fluorosilicic acid is used by checking with a hydrometer.
14. Prepare a metering pump calibration curve for optimum fluoride levels (water flow rate vs. metering pump settings).
15. Test SCADA system to verify a decrease and increase in pacing signal works properly.
16. Keep accurate records of all testing results.

This page intentionally blank.

Index

NOTE: *t.* indicates table; *f.* indicates figure

AWWA Manuals

M1, *Principles of Water Rates, Fees, and Charges,* #30001

M2, *Instrumentation and Control,* #30002

M3, *Safety Management for Water Utilities,* #30003

M4, *Water Fluoridation Principles and Practices,* #30004

M5, *Water Utility Management,* #30005

M6, *Water Meters—Selection, Installation, Testing, and Maintenance,* #30006

M7, *Problem Organisms in Water: Identification and Treatment,* #30007

M9, *Concrete Pressure Pipe,* #30009

M11, *Steel Pipe—A Guide for Design and Installation,* #30011

M12, *Simplified Procedures for Water Examination,* #30012

M14, *Backflow Prevention and Cross-Connection Control: Recommended Practices,* #30014

M17, *Fire Hydrants: Installation, Field Testing, and Maintenance,* #30017

M19, *Emergency Planning for Water Utilities,* #30019

M20, *Water Chlorination/Chloramination Practices and Principles,* #30020

M21, *Groundwater,* #30021

M22, *Sizing Water Service Lines and Meters,* #30022

M23, *PVC Pipe—Design and Installation,* #30023

M24, *Planning for the Distribution of Reclaimed Water,* #30024

M25, *Flexible-Membrane Covers and Linings for Potable-Water Reservoirs,* #30025

M27, *External Corrosion Control for Infrastructure Sustainability,* #30027

M28, *Rehabilitation of Water Mains,* #30028

M29, *Water Utility Capital Financing,* #30029

M30, *Precoat Filtration,* #30030

M31, *Distribution System Requirements for Fire Protection,* #30031

M32, *Computer Modeling of Water Distribution Systems,* #30032

M33, *Flowmeters in Water Supply,* #30033

M36, *Water Audits and Loss Control Programs,* #30036

M37, *Operational Control of Coagulation and Filtration Processes,* #30037

M38, *Electrodialysis and Electrodialysis Reversal,* #30038

M41, *Ductile-Iron Pipe and Fittings,* #30041

M42, *Steel Water-Storage Tanks,* #30042

M44, *Distribution Valves: Selection, Installation, Field Testing, and Maintenance,* #30044

M45, *Fiberglass Pipe Design,* #30045

M46, *Reverse Osmosis and Nanofiltration,* #30046

M47, *Capital Project Delivery,* #30047

M48, *Waterborne Pathogens,* #30048

M49, *Butterfly Valves: Torque, Head Loss, and Cavitation Analysis,* #30049

M50, *Water Resources Planning,* #30050

M51, *Air Valves: Air-Release, Air/Vacuum, and Combination,* #30051

M52, *Water Conservation Programs—A Planning Manual,* #30052

M53, *Microfiltration and Ultrafiltration Membranes for Drinking Water,* #30053

M54, *Developing Rates for Small Systems,* #30054

M55, *PE Pipe—Design and Installation,* #30055

M56, *Nitrification Prevention and Control in Drinking Water,* #30056

M57, *Algae: Source to Treatment,* #30057

M58, *Internal Corrosion Control in Water Distribution Systems,* #30058

M60, *Drought Preparedness and Response,* #30060

M61, *Desalination of Seawater,* #30061

M63, *Aquifer Storage and Recovery,* #30063

M65, *On-Site Generation of Hypochlorite,* #30065

M66, *Actuators and Controls—Design and Installation,* #30066

This page intentionally blank.

www.ingramcontent.com/pod-product-compliance
Lightning Source LLC
LaVergne TN
LVHW082345200926
848481LV00012B/3454
9781625761705